Membrane Shields: Plants Fight Environmental Stress

Class

Table of Contents

Chapter 1. Introduction

Lipids are the basic components of biological membranes, which not only protect the organelles / cells from the outer stimuli (Casares *et al.* 2019, Zajchowski and Robbins 2002), but also display selective permeability towards diverse biomolecules, allowing the exchange and communications between the neighboring cells and between the subcellular organelles within one cell (Record *et al.* 2014, Shinoda 2016). In addition, they are involved in several biological processes, serving as hormonal precursors, signaling molecules and energy sources (Hou *et al.* 2016, Wasternack and Kombrink 2010, Welte and Gould 2017).

Although it seems to be basic knowledge nowadays that the elemental unit of living creatures, the cell, is enclosed by a membrane bilayer, this has only become evident since the second half of the 20th century (Lombard 2014, Lombard *et al.* 2012, Singer 1971, Singer 1974, Singer 2004, Singer and Nicolson 1972). This membrane bilayer, namely the plasma membrane (PM), separates the cell from the exterior environment, thereby guarding the cell from potential harm. As the mobility of plants is restricted, the survival of plants under different biotic and abiotic stresses highly depends on the dynamic responses of the PM according to its environment (Mamode Cassim *et al.* 2019, Mamode Cassim and Mongrand 2019). The alteration of the PM lipid composition not only greatly influences its fluidity and permeability, but also modulates the interactions between membrane components and their downstream signaling pathways (Cai *et al.* 2018, Jiang 2019, Raghunathan and Kenworthy 2018).

The responses and strategies that plants exert to survive exterior hassles are of agronomic and economic interests in terms of increasing food production on non-habitable lands (Bailey-Serres *et al.* 2019, Schroeder *et al.* 2013). Besides pathogenic attacks, freezing and cold stresses are among the most common obstacles that crop plants face (Eremina *et al.* 2016, van Hooren and Munnik 2017, Zhu 2016). It has been demonstrated that low temperature induces drastic changes of the lipid composition in leaf tissues such as increasing the unsaturation degree of lipids and thus the membrane fluidity to avoid freezing injury (Takahashi *et al.* 2018, Tarazona *et al.* 2015, Uemura *et al.* 2006). Noteworthy, sphingolipids, which are critical components of the PM, have been demonstrated to be involved in several signaling pathways including vesicle trafficking, plant development, senescence and defense against biotic and abiotic stresses (Ali *et al.* 2018, Berkey *et al.* 2012, Huby *et al.* 2020, Liang *et al.* 2003, Michaelson *et al.* 2016). In addition, sphingolipids can interact with sterols to establish lipid rafts, which provide specialized lipid platforms within the PM for close interactions between protein receptors and signal transducers (Lefebvre *et al.* 2007,

Takahashi *et al.* 2013, Tapken and Murphy 2015). Nevertheless, the function of lipid rafts during stress adaption and the strategies that plants employ to overcome the loss of specific sphingolipids remained elusive. The role of sphingolipids with respect to the organization of the plant PM and its compositional modulation in response to cold stress are addressed by in-depth profiling of PMs isolated from sphingolipid biosynthesis mutant plants grown under cold acclimation (Chapter 4).

Sphingolipid signaling is involved in regulating programmed cell death (PCD), which may be initiated by processes in the mitochondria, yet only limited information is available in terms of composition and functions of lipids within plant mitochondria (Berkey *et al.* 2012). It is known that the mitochondrion can synthesize some nucleic acids, proteins and lipids on its own; nevertheless, it still strongly relies on the supplies from exterior sources such as the nucleus and the endoplasmic reticulum (ER) (Rhoads and Vanlerberghe 2004, Soto-Heredero *et al.* 2017). Since mitochondria are disassociated from the vesicular transport that utilizes membrane vesicles deriving from the Golgi apparatus, they have developed different strategies to obtain lipids from other organelles such as membrane contact sites (Giordano 2018, Petrungaro and Kornmann 2019). The capacity of lipid biosynthesis of the plant mitochondria and the lipid exchange with other organelles are elucidated here by a combinatorial approach integrating lipidomics, proteomics and online database mining (Chapter 5).

Several methodologies such as thin-layer chromatography (TLC), gas-liquid chromatography (GC) and mass spectrometry (MS)-based approaches have been developed to analyze the broad range of lipids present in biological samples (Carrasco-Pancorbo *et al.* 2009, Pati *et al.* 2016). Liquid chromatography – mass spectrometry (LC-MS) is the most sensitive and efficient approach hereto in identifying single molecular lipid species in biological extracts. A wide-ranging lipidomics platform that covers more than 300 molecular lipid species from *Arabidopsis* leaves has been developed previously (Tarazona *et al.* 2015). This lipidomics platform empowers greatly the research on lipid characterization. However, some lipid classes that exert regulatory functions are challenging to be incorporated into the standard workflow due to their low abundance and / or structural characteristics. For instance, phosphoinositides and phosphorylated sphingolipids, which play essential roles in regulating several signaling pathways and in determining cell fate, respectively, were not included in the pre-existing lipidomics platform. Our strategy to include the minor but functional lipid classes, enhance the detection and broaden the coverage of the pre-existing lipidomics platform is delineated in detail (Chapter 2 and 3).

1.1 Definition, classification and nomenclature of lipids

In this study, lipids are defined according to the International Union of Pure and Applied Chemistry (IUPAC-IUB Commission on Biochemical Nomenclature) and the LIPID MAPS Consortium (Fahy *et al.* 2011, Harkewicz and Dennis 2011). They are hydrophobic or amphipathic chemical compounds that originate entirely or partially from carbanion-based condensations of thioesters (such as glycerolipids and sphingolipids) and/or by carbocation-based condensations of isoprene units (sterols). The three major lipid categories, namely glycerolipids, sphingolipids and sterols, are within the focus of this work (Fig. 1). Within each lipid category, individual lipid classes are named purely after their structural backbones and functional groups. A three-number system (C : DB ; OH) connected by a colon and a semicolon is used here to denote the numbers of carbon (C), double bond (DB) and hydroxyl group (OH) of the hydrocarbon moieties (Liebisch *et al.* 2013). Only the first two digits are specified if no hydroxyl group is present. For instance, 18:0;0 and 18:0 fatty acids represent the same lipid structure with an 18-carbon acyl chain without any double bond or hydroxyl group.

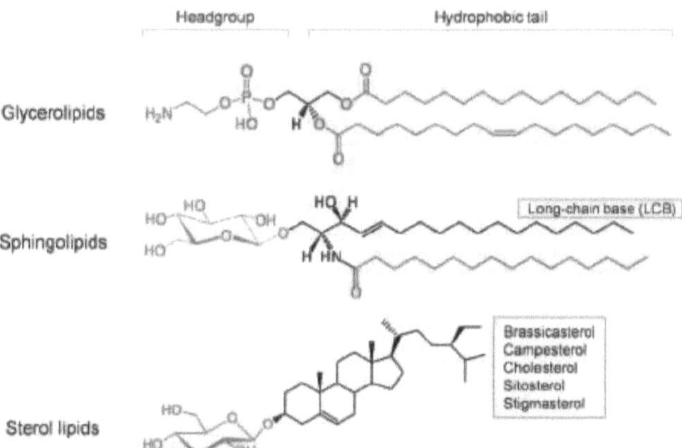

Figure 1. General structures of glycerolipids, sphingolipids and sterols. The lipid categories are named after the core structure (glycerol for glycerolipids, LCB for sphingolipids, tetracyclic ring for sterol lipids) depicted in black. The lipid classes within each lipid category vary according to the headgroups depicted in blue. Lipid species within each lipid class are specified according to their different fatty acyl moieties (glycerolipids and sphingolipids) or side-chain modifications on the core structure (sphingolipids and sterol lipids) are depicted in magenta.

Glycerolipids are a group of structurally heterogeneous compounds that have glycerol as their backbone. The central glycerol, which contains three carbon atoms that are numbered stereospecifically with *sn*-1, *sn*-2 and *sn*-3, can be linked with at least one fatty acyl moiety. The *de novo* biosynthesis of the glycerolipids is initiated in plants in the plastids where fatty acid moieties with 16 and 18 carbons in length are generated. These fatty acyl moieties can then be transported to the plastidial envelope or exported to the ER for further desaturation, elongation and incorporation into a varieties of lipid molecules. Substantial amounts of the lipids may be transferred back into plastids as additional precursors for the synthesis of plastidial lipids (Kelly and Dörmann 2004, Li-Beisson *et al.* 2013). According to the involved compartments of the biosynthesis pathways, the prokaryotic and the eukaryotic pathways are defined (Browse *et al.* 1986). Namely, the prokaryotic pathway takes place exclusively in plastids, and the resulting prokaryotic lipids characteristically contain in *Arabidopsis* fatty acyl moieties with 16 carbons in length (16:3 in particular) at the *sn*-2 position. Most of the glycerolipid biosynthesis in *Arabidopsis* follows the eukaryotic pathway. It involves both plastids and the ER and the resulting eukaryotic lipids contain primarily fatty acid moieties with 18 or more carbons in length.

The simple examples of the glycerolipids with one, two and three acyl chains are the mono-, di- and triacylglycerols (MAG, DAG and TAG). In plants, DAG serves as an important signaling molecule during plant development and in response to environmental stresses (Dong *et al.* 2012); TAG is the predominant lipid component in seeds, providing energy as well as acting as carbon source for the young seedlings during early germination stages (Huang 1994, Lu *et al.* 2020). Glycerolipids can be further subdivided into glycerophospholipids and glyceroglycolipids, which contain a phosphate group or a carbohydrate attached to the *sn*-3 position of the glycerol core, respectively.

Most of the glycerolipids are glycerophospholipids, which are amphiphilic and play central roles in forming lipid bilayers. The glycerophospholipid phosphatidic acid (PA) is the simplest phospholipid and is formed after serial acylation of the glycerol-3-phosphate (Athenstaedt and Daum 1999, Testerink and Munnik 2011). The phosphate group on the *sn*-3 position can be modified further with small molecules such as choline, ethanolamine, serine, inositol and glycerol to generate the following lipid classes: phosphatidylcholine (PC), phosphatidylethanolamine (PE), phosphatidylserine (PS), phosphatidylinositol (PI) and phosphatidylglycerol (PG), respectively. Although glycerophospholipids like PC and PE are widely recognized as the structural components of biological membranes, many of them play essential roles in several biological processes as well (Gibellini and Smith 2010, Hong *et al.* 2009, Karki *et al.* 2019). In plants, PS acts as an early indicator of apoptosis (Manoharan *et al.* 2000, O'Brien *et al.* 1997). PI is the precursor of phosphatidylinositol mono- and bisphosphates (PIP and PIP$_2$), which form a sophisticated signaling network in regulating the location and the activities of specific proteins (Heilmann 2016, Thole and Nielsen 2008). Noteworthy, glycerophospholipids may be also involved in the determination of membrane curvature and surface charge. For instance, a negative curvature can be induced by PA and PE, and a positive curvature by PI, PIP and PIP_2 (Harayama and Riezman 2018). The surface charge of the membrane, which establishes its electrostatic signature, depends on the presenting anionic lipids including PA, PS, PIP and PIP_2 (Platre *et al.* 2018). PG is mainly present in the thylakoid membranes to support the performance of photosynthesis (Hagio *et al.* 2002, Wada and Murata 2007). In mitochondria, PG can serve as the precursor of the mitochondria-specific phospholipid, cardiolipin (CL) (Pineau *et al.* 2013). Unlike other lipid classes, the majority of PG (85 %) is produced entirely in the plastid via the prokaryotic pathway (Browse *et al.* 1986). Three common glyceroglycolipids in plants are monogalactosyl, digalactosyl and sulfoquinovosyl diacylglycerol (MGDG, DGDG and SQDG, respectively). They are critical components of the thylakoid membranes in higher plants and are involved in the photosynthesis process (Kobayashi *et al.* 2007, Li and Yu 2018). In *Arabidopsis*, about 50 % of MGDG is of prokaryotic origin, whereas DGDG and SQDG are primarily of eukaryotic origin (Li-Beisson *et al.* 2013). Noteworthy, DGDG is transferred to extraplastidial membranes including the PM and mitochondrial membranes under phosphate-limiting condition (Jouhet *et al.* 2004, Kelly and Dörmann 2004, Michaud *et al.* 2016). SQDG levels correlate strongly with plant growth under phosphate-limited stress as well (Benning *et al.* 2008, Shimojima 2011). Further lipid classes such as betaine lipids and complex glycolipids with tri- or tetragalactosyl headgroups are widely distributed in algae and lower plants (Eichenberger *et al.* 1993, Murakami *et al.* 2018, Vogel and Eichenberger 1992, Warakanont *et al.* 2015).

Sphingolipids, originally discovered in brain tissues, are ubiquitous structural and functional components of eukaryotic membranes (Luttgeharm *et al.* 2016). They are comprised of a group of structurally diverse lipids that have a long-chain base (LCB) as the consensus core. These LCB molecules, which are synthesized in the ER, are long-chain amines that contain two or three hydroxyl groups (Chen *et al.* 2008). *N*-acylation of the LCB backbone with a fatty acyl moiety generates ceramide (Cer), and subsequent glucosylation generates glucosylceramide (GlcCer). Cer can be transported to the Golgi apparatus for synthesizing complex sphingolipids such as series of glycosyl inositol phosphoceramides (GIPCs). According to the nomenclature proposed previously (Fang *et al.* 2016), series 0 GIPC in plants, which usually contains an α-glucuronic acid linked to the inositol phosphoceramide, can be denoted as GlcA-IPC. Subsequent addition of one and two sugars on the α-glucuronic acid generates series A and B GIPCs, respectively. Sphingolipids play roles in numerous biological processes. For instance, the ratio between the non- and phosphorylated forms of LCB (LCB/LCB-P) and Cer (Cer/Cer-P) has been demonstrated to be a decisive factor of cell fate determination including senescence and PCD (Huby *et al.* 2020, Luttgeharm *et al.* 2016). On the other hand, GlcCer is involved in the processes of cell differentiation and organogenesis (Msanne *et al.* 2015, Warnecke and Heinz 2003). It has been proposed to establish lipid rafts by interacting with sterols and specific proteins as well (Quinn 2014, Varela *et al.* 2016). GIPCs carry different glycan moieties on the head groups and are mainly located at the PM (Cacas *et al.* 2016, Gronnier *et al.* 2016). In *Arabidopsis*, GIPCs are required for salt-induced depolarization (Jiang *et al.* 2019) and pathogenic recognition (Lenarčič *et al.* 2017, Mortimer and Scheller 2020).

Sterols are the third lipid category included in this study. Unlike animal and yeast, which have predominantly cholesterol and ergosterol, respectively, a mixture of free sterols (FSs) are present in plants. In the model plant *Arabidopsis* it is a mixture out of brassicasterol, campesterol, cholesterol, isofucosterol, sitosterol and stigmasterol (Clouse 2002, Schaller 2004). All the mentioned sterols share the same backbone, a tetracyclic ring system, but differ in the additional modifications on the side chain and the four ring systems. FSs can be further conjugated via a hydroxyl group at position 3 of the A ring with fatty acids via an ester bond to generate sterol esters (SEs), or with glucose via glycosidic linkage to generate steryl glycosides (SGs) and further acylated steryl glycosides (ASGs) (Hartmann 2004, Valitova *et al.* 2016). It is well-characterized that FSs are modulators of fluidity and permeability of biological membranes (Grunwald 1971). SE molecules are considered as storage forms of sterols and accumulate in lipid droplets (Bouvier-Nave *et al.* 2010). On the other hand, SGs and ASGs, which present specifically in plants, are less characterized. It is proposed that SGs and ASGs, together with FSs are associated with GlcCers and GIPCs to form lipid rafts (Ferrer *et al.* 2017, Grosjean *et al.* 2015, Roche *et al.* 2008).

1.2 Composition and organization of biological membranes

Although glycerolipids, sphingolipids and sterols are structurally and functionally different, one common role that all three lipid categories exert is to establish biological membranes. Current knowledge shows that the biological membranes in eukaryotic systems are generated from the ER (Balla *et al.* 2020, Henneberry *et al.* 2002, Kent 1995, Yang *et al.* 2018). The majority of the membrane lipids are synthesized here and then transported by active mechanisms, such as the vesicular transport, to the targeted destinations. Lipids can undergo subsequent modification such as double bond insertion, chain length elongation and glycosylation after being synthesized in the ER, as well as during the transport through the Golgi apparatus and / or other endomembranes (Balla *et al.* 2020, Stefan *et al.* 2017). Together with lipids, the membrane proteins, which are synthesized in the ER, are transported via the vesicular transport as well. Based on the lipid – protein interactions, these proteins form specialized clusters with the surrounding lipids, and are incorporated together into the acceptor compartment (Brown 2012, Quinn 2012). Thereby, the interactions between lipid – lipid, protein – protein and lipid – protein on the biological membranes collectively determine the biophysical properties of the membrane, building a dynamic but stable platform for numerous biological processes.

Membrane lipids and the associated proteins are distributed asymmetrically not only across the two leaflets of the membrane, but also laterally in each leaflet in a non-random manner (Devaux and Morris 2004, Fujimoto and Parmryd 2017). It is proposed that the transmembrane asymmetry results from the nature of membrane lipid biosynthesis and transport (Balla *et al.* 2020); and the lateral membrane asymmetry is organized majorly by the lipid – lipid and lipid – protein interactions (Quinn 2012). One example that drastically affects the architecture of the membrane due to the ordered lipid – protein structure are the membrane junctions. In this case, proteins are the central scaffold of the membrane junctions that impose the type and distribution of the surrounding lipids (Okeke *et al.* 2016, Van Itallie and Anderson 2014). Concerning lipid – lipid interaction, the most notable example is the formation of lipid rafts. They have been characterized as ordered lipid structures that gather membrane signaling components and facilitate downstream signaling responses (Grennan 2007, Hanzal-Bayer and Hancock 2007).

The definition and existence of lipid rafts have been long under debate (Brown 2006). Historically, liquid-ordered membranes isolated after treatment with mild detergent, namely the detergent-resistant membranes (DRMs), were used to study the naturally organized membrane structures present on biological membranes (Borner *et al.* 2005, Magee and Parmryd 2003). DRMs contain higher abundance of sterols and sphingolipids in comparison to the PM crude extracts. Furthermore, many studies have demonstrated that a specific subset of proteins can be coisolated with DRMs, suggesting that these proteins may play functional roles in lipid rafts (Lefebvre *et al.* 2007, Magee and Parmryd 2003). Although many scientists raised the concern that the DRMs and the natural lipid rafts are biophysically different (Lichtenberg *et al.* 2005), several DRM-associated proteins and lipids have already been visualized by microscopic techniques to cluster towards specific receptors within the PMs (Gaus *et al.* 2005, Viola *et al.* 1999). However, the majority of the most typical DRM-associated proteins, the glycosyl phosphatidylinositol-anchored proteins (GPI-AP), distribute uniformly or in nanoscale clusters across the cell surface (Glebov and Nichols 2004, Sharma *et al.* 2004). A revised model has been therefore proposed. Namely, lipid molecules are present uniformly or in nanoscale unstable domains in the PM under the unstimulated condition; upon stimulus, lipids such as sterols and sphingolipids initiate the formation and increase the stability of larger lipid rafts (Carquin *et al.* 2016, Cebecauer *et al.* 2018). This concept has been demonstrated through an *in vitro* study with PM-like giant unilamellar vesicles (GUVs) (Hammond *et al.* 2005). The composition of these GUVs represents the PM condition, which is at the boundary of phase separation. The initial aggregation of lipid or protein molecules can generate a large impact on their membrane organization to form lipid-ordered domains or lipid rafts. In the present work, lipid rafts are defined as following: they are lipid microdomains that contain ordered structures of lipid molecules including sterols and sphingolipids; they harbor a specific subset of proteins, usually receptors or signaling molecules; they may be involved in mediating or facilitating the signaling transduction, thereby transmitting the exterior stimuli into the cell.

1.3 Targeted membranes in this study

Two distinct membranes are in the scope of this work, the PM and the mitochondrial membranes in the model plant *Arabidopsis*. These two membrane systems represent the eukaryotic and prokaryotic membranes in plant cells, and both are pivotal for survival. The following sections focus on the function, structure, biosynthesis and the composition of PM and mitochondria.

1.3.1 Plasma membrane

The PM is the frontier of the cell that encloses and separates the cytoplasm from the exterior environment while ensuring the coordination between internal responses and the external stimuli. The basic PM structure is a phospholipid bilayer with additional lipids such as sterols for increasing stability, and sphingolipids for mediating numerous biological processes. There are also several PM-associated proteins, such as integral, peripheral and lipid-anchored proteins (Chou and Elrod 1999). Furthermore, the PM surface possesses various forms of carbohydrates, predominantly from the glycoproteins and glycolipids. They play critical roles in mediating cell attachment, as the recognition sites between neighboring cells, as well as between host cells and pathogens (Chaliha *et al.* 2018, Kieliszewski *et al.* 2011).

For the PM, which lacks the synthetic enzyme machineries for *de novo* biosynthesis, it obtains the lipids primarily via vesicular transport to maintain its membrane structure (Bishop and Bell 1988, Blom *et al.* 2011, van Meer *et al.* 2008). The vesicular transport involves the budding of vesicles from the donor membrane, transport of the vesicles and fusion with the acceptor membrane (Fig. 2). GIPCs, for instance, are transported via this pathway to the PM. GIPCs are considered to be synthesized on the luminal side of the Golgi apparatus where glycosylation occurs, transported via Golgi-derived vesicles within the cell and delivered to the PM and then exposed to the apoplastic leaflet via an active transport (van Meer and Holthuis 2000, Zäuner *et al.* 2010).

Because the PM possesses different function and structure as the ER and the Golgi apparatus, specific proteins and lipid molecules must be carefully selected to be transported to the PM (Gronnier *et al.* 2018, Li *et al.* 2020). To achieve precise sorting and transport of PM-localized proteins, specific recognition sequences are used; however, there are no recognition sequences for lipids. Therefore, to achieve accurate sorting of PM-specific lipids, the vesicular transport must be coordinated with other mechanisms such as the protein-facilitated non-vesicular pathway and transport at the membrane contact sites when two membranes are at close appositions (Balla *et al.* 2020, Prinz 2010).

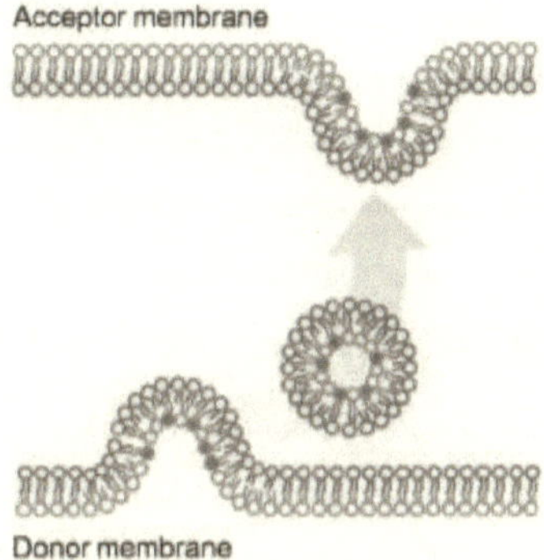

Figure 2. The orientation of lipid molecules at membrane vesicles during the vesicular transportation. Lipid molecules located at the inner membrane of the donor membrane (depicted in gray), such as the luminal side of the ER or the Golgi, are enclosed in the inner leaflet of the membrane vesicles during transportation. In case of membrane vesicles that travel to the PM (acceptor membrane), the targeted lipid molecules (magenta) are exposed at the outer leaflet after membrane fusion.

It has been demonstrated that the PM lipid composition depends highly on the plant species, organs and its growth condition. Under optimal growth condition, the proportion of glycerolipids, sphingolipids and sterols in the PM isolated from leaf tissues of oat, rye, barley, spinach, potato and *Arabidopsis* varies greatly. In most of the mentioned plant species, glycerophospholipids are the most abundant lipid classes in the PM, accounting for 42 % in oat, 44 % in barley (Rochester *et al.* 1987), 47 % in *Arabidopsis* (Uemura *et al.* 1995), 48 % in rye (Takahashi *et al.* 2016), 47-50 % in potatoes (Palta *et al.* 1993) and 64 % in spinach (Rochester *et al.* 1987). Interestingly, in spring and winter cultivars of oat and rye PM, significant lower abundance of glycerophospholipids and high levels of sterols are present. Namely, 30 % : 42 % (glycerophospholipids : sterols) in spring oat, 29 % : 40 % in winter oats and 37 % : 47 % in winter rye (Uemura and Steponkus 1994). This demonstrates not only how evolution has shaped the PM lipid composition of each plant species but also to which extent the lipid composition of the plant PM can adapt to outer stimuli while maintaining its membrane stability. Sphingolipids, on the other hand, are minor components in the plant PM. They account for only 7 % in the PM isolated from leaves of *Arabidopsis* (Uemura *et al.* 1995) although they exert essential functional roles including lipid raft formation and the pathogenic recognition.

1.3.1.1 Cold acclimation

Cold and freezing stresses are among the major obstacles that affect the growth and development of plants. It has been demonstrated that changes in the lipid and protein compositions of the plant PM, and alterations of the overall transcriptome and proteome inside the cell arise when plants are exposed to cold stress (Chinnusamy *et al.* 2006, Nakashima and Yamaguchi-Shinozaki 2006, Uemura *et al.* 2006, Van Buskirk and Thomashow 2006). To maintain the homeostasis of the cell, plants have evolved a strategy, namely cold acclimation (CA), to prepare themselves already at low temperature for the upcoming freezing stress.

The most typical strategy when plants encounter cold stress is to increase the cryostability and maintain the fluidity of the PM by modulating its lipid composition. At the early stage of CA, increasing proportions of phospholipids have been observed commonly (Uemura *et al.* 2006). The molecular species of PC and PE change as follows: the levels of the unsaturated species (containing unsaturated fatty acyl moieties) increase and saturated species (containing saturated fatty acyl moieties) decrease (Uemura *et al.* 1995). In addition, levels of sphingolipids decrease continuously throughout the CA process. More unsaturated GlcCers are present under CA condition in oat, rye, cereal and *Arabidopsis* (Huby *et al.* 2020, Imai *et al.* 1997, Minami *et al.* 2008, Takahashi *et al.* 2016). The increase of the unsaturated sphingolipids probably contributes to the structural stability of the plant PM as well. However, the alteration of the phosphorylated classes including LCB-Ps and Cer-Ps initiates presumably signaling pathways adjusting the response or the degradation of the precursors LCB and Cer from the membranes (Chen *et al.* 2012, Dutilleul *et al.* 2012, Dutilleul *et al.* 2015, Michaelson *et al.* 2016). In addition, plant sterols, including the free forms and the glycosylated conjugates, are important modulators of the membrane fluidity. For instance, ratio of FSs to SGs and ASGs increase in rye and *Arabidopsis* during CA (Grosjean *et al.* 2015, Lynch and Steponkus 1987, Minami *et al.* 2008). As SGs, ASGs and sphingolipids are considered as critical components of the lipid rafts, the change of their molecular species affects the involving biological processes such as signal transduction, protein trafficking and plant – bacterial interactions (Bhat and Panstruga 2005, Haney and Long 2010, Lefebvre *et al.* 2010, Men *et al.* 2008, Minami *et al.* 2008, Tapken and Murphy 2015, Yang *et al.* 2013).

In addition to lipids, the composition of proteins and their interactions change during CA as well (Takahashi *et al.* 2013). Several PM proteins which are involved in membrane repairing, osmotic stress responses and protein degradation have been identified by proteomics approaches to take part in the CA (Kawamura and Uemura 2003). Moreover, it has been demonstrated that a lipoprotein-like protein, AtLCN, accumulates significantly during CA, and may be involved in increasing the cryostability of the PM by interacting with the neighboring lipid components (Kawamura and Uemura 2003, Uemura *et al.* 2006). Furthermore, many other lipid raft-localized or associated proteins accumulate in the PM during CA, such as tubulins, clathrins and aquaporins (Abdrakhamanova *et al.* 2003, Minami *et al.* 2008, Peng *et al.* 2008, Takahashi *et al.* 2013), suggesting that the protein – lipid interactions play important roles during CA as well.

1.3.1.2 Sphingolipid synthesis defects

Sphingolipids are not only structural but also functional components of the plant PM. In *Arabidopsis*, the LCB backbones of sphingolipids are usually 18 carbons in length, with maximum two double bonds and three hydroxyl groups. The amide-linked fatty acyl moieties on complex sphingolipids vary in chain length usually from 16 to 26 carbons, with possible modifications such as the hydroxylation at the C2 position (α-hydroxylation) and the desaturation at the ω-9 position (Alderson *et al.* 2005, Imai *et al.* 2000, Lynch and Dunn 2004, Michaelson *et al.* 2016).

More than 90 % of the complex sphingolipids in higher plants are α-hydroxylated (Imai *et al.* 1995, Markham and Jaworski 2007, Pata *et al.* 2010). It has been demonstrated that α-hydroxylated sphingolipids are involved in several biological processes. For instance, hydroxylated complex sphingolipids, especially the ones containing very-long-chain fatty acids (VLCFAs), are associated with sterols in establishing and stabilizing lipid rafts (Borner *et al.* 2005, Cacas *et al.* 2016). In addition, plants which contain high levels of hydroxylated Cer display enhanced susceptibility upon hypoxia (Li *et al.* 2015, Xie *et al.* 2015a, Xie *et al.* 2015b) whereas reduced abundance of hydroxylated Cer has been detected under active ethylene signaling (Wu *et al.* 2015), suggesting its functions in oxidative stress and cell development.

Two sphingolipid fatty acid α-hydroxylases (FAH), AtFAH1 and AtFAH2, have been identified in *Arabidopsis* to synthesize α-hydroxylated sphingolipids. Although AtFAH1 and AtFAH2 share very high sequence similarity, they possess different substrate specificities. AtFAH1 is capable of hydrolyzing a broad range of substrates, preferentially the VLCFAs, while AtFAH2 reacts specifically on the 16:0 fatty acid (König *et al.* 2012, Nagano *et al.* 2012). Hydroxylated VLCFAs have been demonstrated to be involved in enhancing the tolerance to oxidative stress and salicylic acid (SA)-

triggered cell death. In addition, AtFAH1 and AtFAH2 have been identified to interact with and mediate the functions of several ER-localized proteins. For instance, both FAH proteins interact with the Bax inhibitor-1 (AtBI-1), which function as a cell death suppressor (Nagano *et al.* 2009). Noteworthy, AtBI-1 overexpressing plants contain higher amount of hydroxylated GlcCer in the lipid rafts with altered protein composition (Ishikawa *et al.* 2015), indicating again the correlation between lipid rafts and stress responses.

The *Arabidopsis fah1 fah2* double mutant displays reduced growth phenotype under optimal growth condition. Further analysis of overall lipid composition in leaves of *fah1 fah2* indicated that they contain higher amounts of free trihydroxylated LCB, C16- and VLCFA-containing Cers; but lower amount of hydroxylated complex sphingolipids (König *et al.* 2012). Constitutively enhanced levels of SA and SA derivatives, together with other stress markers such as raphanusamic acid, indoles and dihydroxybenzoic acid derivatives, have been detected as well (Bartsch *et al.* 2010, Bednarek and Osbourn 2009, Hagemeier *et al.* 2001, Iven *et al.* 2012). Collectively, the *fah1 fah2* double mutant serves as a great model to investigate the influence of sphingolipids, especially hydroxylated complex sphingolipids, on the organization of the plant PM under environmental stresses.

1.3.1.3 Membrane asymmetry

It has been widely accepted that the PM of eukaryotic cells contains non-randomly distributed lipids across the two leaflets of the membrane, namely lipid asymmetry. The uneven distribution of lipid molecules contribute to the establishment of the curvature and the charge of the membrane (Bigay and Antonny 2012, Harayama and Riezman 2018). Most of the knowledge about PM lipid asymmetry is building on studies of human erythrocytes. That is, most of the PC and sphingomyelin (SM) are distributed in the outer leaflet while PS, PE and PI are in the inner leaflet (Bretscher 1972, Devaux and Morris 2004, Lorent *et al.* 2020, Verkleij *et al.* 1973). It was assumed that this PM prototype could be applied generically on other cell types and species. However, several studies indicate that the transversal lipid distribution varies among cell types and even among erythrocytes of different species. For instance, the proportion of PC in the outer leaflet of erythrocytes is about 77 % in human (Verkleij *et al.* 1973), but only 50 % in mouse (Rawyler *et al.* 1985). PS, PE and PI are thought to locate majorly in the inner leaflet of human erythrocytes. However, considerable amounts of PS and PE have also been detected by lipid probes, although not quantitatively, in the outer leaflet of animal cells (Devaux 1991, Zachowski 1993).

In plants, the knowledge of the lipid transversal distribution is still in its infancy; very few studies have addressed it experimentally. It has been demonstrated that phospholipids are allocated symmetrically on the two leaflets of the PM isolated from hypocotyl cells of mung beans (Takeda and Kasamo 2001); 65 % of the overall phospholipids are located in the cytoplasmic leaflet in the PM isolated from oat roots, and the cytoplasmic leaflet-localized PC can be replaced by DGDG under phosphate starvation (Tjellström *et al.* 2010). With the assistance of protein-based lipid probes, the transversal distribution of anionic lipid classes was further revealed. PS is detectable preferentially, but not exclusively, in the cytoplasmic leaflet of tobacco protoplasts by the PS probe, Annexin-V (O'Brien *et al.* 1997); together with PS, PA and PIPs are detected with higher abundance in the cytoplasmic leaflet of *Arabidopsis* root meristem (Platre *et al.* 2018).

Plants possess a wide range of sphingolipids and sterols in comparison to the animal system. Nevertheless, the knowledge obtained from the human erythrocyte serves as a foundation for further investigations. It has been widely accepted that glycosphingolipids are localized exclusively on the apoplastic leaflet of the PM according to their biosynthesis and transport pathway. Since the complex sugar moieties are attached to the molecules on the luminal side of the Golgi apparatus and this hydrophilic headgroup can hinder the transbilayer movement, the glycosphingolipids are therefore exclusively exposed on the apoplastic leaflet of the PM after membrane fusion. However, it has been demonstrated that monohexosyl sphingolipids undergo rapid transbilayer movement on the ER and the Golgi membranes of rat liver (Buton *et al.* 2002). This suggests the presence of specific flippases acting on membranes of the ER and the Golgi apparatus, and may contribute to the presence of glycosylated sphingolipids within the inner leaflet of the PM. In plants, although a wide variety of glycosphingolipids are present, only GlcCer have been investigated. The proportion of GlcCer located within the apoplastic leaflet of the PM isolated from summer squash and from roots of oat was reported to be 98 % (Lynch and Phinney 1995) and 70 % (Tjellström *et al.* 2010), respectively.

Analyses of the sterol distribution have demonstrated that sterols are located, not randomly, to either monolayer of the PM. In mammalian systems, the proportion of cholesterol varies greatly among the cell types and the applied methodology (Fujimoto and Parmryd 2017): 60-70 % are localized to the inner leaflet of the PM from Chinese hamster ovary cells by fluorescent analysis (Mondal *et al.* 2009), but 50-74 % to the outer leaflet of the PM from human erythrocyte by freeze-fracturing (Fisher 1976). In plants, a sterol probe, Filipin III (Kleinschmidt *et al.* 1972), has been applied on targeting FSs and sterol derivatives thus addressing the sterol distribution. In PM isolated from roots of oat, 70 % of the overall sterols are present in the apoplastic leaflet (Tjellström *et al.* 2010).

In addition to the functional headgroups, the unsaturation degree of lipid molecular species affects its transmembrane distribution as well. It has been demonstrated that the inner leaflet of the PM isolated from animal cells possesses a higher unsaturation degree (Devaux 1991, Lorent *et al.* 2020), lower viscosity (Morrot *et al.* 1986) and higher mobility of lipid molecules (Julien *et al.* 1993) in comparison to the outer leaflet. Moreover, increasing evidence suggests that individual lipid species can exert important roles in distinct cellular processes such as signal transduction and cell development. However, the lipid asymmetry of the plant PM in the scope of molecular species remains to be determined.

1.3.2 Mitochondria

The mitochondrion is enclosed by two biological membranes, the inner and outer mitochondrial membranes (IM and OM), and it exists in all kinds of eukaryotic cells. Knowledge about the mitochondrion is majorly building on the research of mammalian cells and yeast. The OM contains several protein channels for importing nucleotides, peptides, metabolites and ions into the inter membrane space. The second membrane, the IM, forms compartments by folding the membranes into specialized structures, the cristae. The complex machinery of electron transport chain is embedded in the IM, and the cristae are required for its efficient energy generation. The two aqueous compartments in mitochondria, the intermembrane space (between OM and IM) and the matrix (enclosed by IM), play essential roles in the regulation of electron transport chain and the biosynthesis of mitochondrial proteins.

Noteworthy, mitochondria in plants have specialized during the evolution according to the tissue and cell type. In plants, which lack mobility, mitochondria are obligated to cooperate with other organelles including plastids under different environmental stresses in order to support plant growth and development. One of the most important cellular events triggered by mitochondria is PCD (Hirsch *et al.* 1998). It has been demonstrated that PCD can be triggered by the release of cytochrome *c* from the mitochondria and be inhibited by BCL-2 protein in mammalian cells (Kluck *et al.* 1997, Yang *et al.* 1997) and in plants (Chen and Dickman 2004, Dion *et al.* 1997). Interestingly, immunoblotting analysis indicates that the plant BCL-2 protein is associated not only with mitochondria, but also with plastids and nuclei, suggesting that PCD may be regulated by multi-organellar signals in plant cells. Interactions between mitochondria – nucleus, mitochondria – ER and mitochondria – plastids further indicate that mitochondria cooperate with other organelles and exert functions in a sophisticated network (Carrie *et al.* 2013, Mackenzie and McIntosh 1999).

Several researches have been devoted to the understanding of mitochondrial proteins, including the biosynthesis of proteins inside mitochondria as well as the import of mitochondrial proteins from the cytosol through OM and IM (Wiedemann and Pfanner 2017). However, knowledge of mitochondrial lipids is much less developed. Although the mitochondrion can produce some lipids, it is still necessary to import several lipids from other organelles, especially the ER, during membrane genesis (Michaud *et al.* 2017). Unlike the PM, which obtains lipids from vesicles composed of specific protein and lipid molecules, mitochondria are disassociated from this vesicular transport. Instead, mitochondria obtain lipids from exterior sources mainly via the non-vesicular lipid transport mediated by lipid-transfer proteins or spontaneous lipid transport. The efficiency of the non-vesicular transport is enhanced at the membrane contact sites; that is, when two membranes come to close apposition or even form dynamic membrane bridges (Lev 2010, Prinz 2010). Several morphological and biochemical studies have characterized the membrane contact sites between the mitochondria and the ER, the mitochondrion-associated membranes (MAMs). It is a specialized region where the ER and the OM of the mitochondria are tightly connected. Many enzymes involved in the biosynthesis of phospholipids such as PC, PE and PS display high activities at the purified MAMs from yeast (Vance 1990). In yeast and mammalian systems, it has been demonstrated that several phospholipids are transferred through the MAMs from the ER to the mitochondrial membranes (Herrera-Cruz and Simmen 2017, Tatsuta *et al.* 2014). As MAMs have been identified ubiquitously among animal and plant cells (Achleitner *et al.* 1999, Michaud *et al.* 2016, Morré *et al.* 1971, Staehelin 1997), it is presumed that the plant mitochondria obtain lipids from the ER via the MAMs as well.

The mitochondrial membrane contains primarily phospholipids with minor amounts of sphingolipids and sterols under optimal growth condition (Daum and Vance 1997). It is proposed that plant mitochondrion can produce PE, PA, PG and CL on its own (Flis and Daum 2013, Horvath and Daum 2013, Tatsuta *et al.* 2014) and obtains other lipids from other intracellular organelles such as the ER and the plastids (Fig. 3). Noteworthy, CLs are a group of mitochondrion-specific phospholipids that contain four acyl chains and two phosphate groups. They are essential in establishing the cristae and maintaining the organization of the electron transport chain (Pineau *et al.* 2013). CL is formed after the condensation of PG and CDP-DAG by CL synthase (CLS). In plants, the PG-synthesizing enzymes have been identified to associate with the ER, the plastids and the IM, whereas CLS is identified exclusively in the mitochondria (Katayama *et al.* 2004, Li-Beisson *et al.* 2013, Nowicki *et al.* 2005, Xu *et al.* 2002).

Mitochondria obtain several lipids from other organelles through the non-vesicular transport. For instance, they obtain PC, PE and PI from the ER or the MAMs, and are able to self-synthesize considerable proportions of PS and PG (Michaud *et al.* 2017, van Meer *et al.* 2008). Noteworthy, it has been demonstrated that the plant mitochondria contain the typical plastidial lipids such as MGDG and DGDG as well (Jouhet *et al.* 2004, Kelly and Dörmann 2004). Under phosphate starvation, the mitochondria obtain significant amounts of the bilayer-forming DGDG via the specific mitochondrial transmembrane lipoprotein (MTL) complex (Jouhet *et al.* 2019, Jouhet *et al.* 2004, Michaud *et al.* 2016, Michaud and Jouhet 2019). The increase of DGDG is accompanied by the decrease of PC (also bilayer-forming) and PE in the mitochondrial membranes, which are supposed to be degraded to release the phosphate residues for other essential biological processes. The mitochondrial lipid marker, CL, has increased abundance under phosphate starvation as well (Jouhet *et al.* 2004). Noteworthy, this drastic lipid remodeling and alteration of membrane composition does not affect the general structure and function of mitochondria. However, further knowledge based on the detailed lipid composition including glycerolipids, sphingolipids and sterols of the mitochondria as well as their capacity of the lipid biosynthesis is required to understand its underlying mechanism.

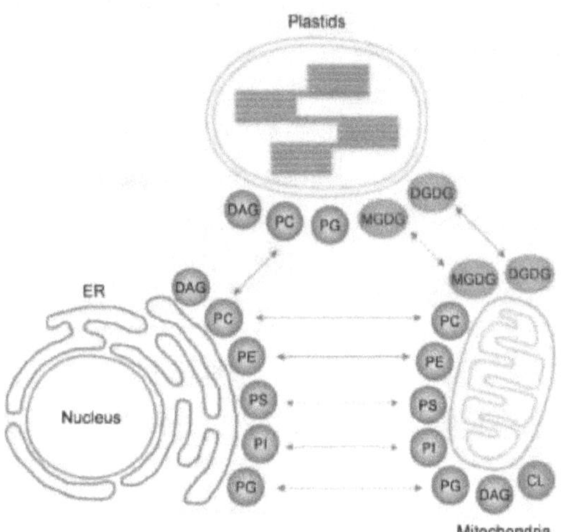

Figure 3. The lipid transportation between the ER, the plastids and the mitochondria. The mitochondrial lipids including PC, PE, PS, PI, PG, and DAG are synthesized in the ER, the MAM or the mitochondria (purple). CL is present exclusively in mitochondria (magenta). Glyceroglycolipids including MGDG and DGDG are transferred from the plastids to the mitochondria (green), with increasing abundance under phosphate starvation. Solid line: exchange of lipids at the membrane contact sites between the organelles; dashed line: the transportation mechanism is still elusive. CL: cardiolipin; DAG: diacylglycerol; DGDG: digalactosyldiacylglycerol; MGDG: monogalactosyldiacylglycerol; PC: phosphatidylcholine; PE: phosphatidylethanolamine; PG: phosphatidylglycerol; PI: phosphatidylinositol; PS: phosphatidylserine.

1.4 Lipid methodology

Detailed knowledge of lipid composition is part of the fundamental information to understand the structure, function and regulation of a membrane. Much effort has been made to develop methods that extract the information of lipid class and lipid species from biological samples, ranging from basic chromatographic separation to the advanced mass spectrometry-based analyses. Three major approaches are broadly applied to identify qualitatively and / or determine quantitatively the biological lipids. That is, (1) the conventional chromatographic analysis, (2) the direct infusion – mass spectrometry and (3) the liquid chromatography – mass spectrometry. The following sections introduce the advantages, disadvantages and the applications of these methodologies.

1.4.1 Conventional chromatographic analysis

The conventional approaches of lipid analysis usually involve two essential stages: the separation of specific lipid classes from complex biological extracts and the detection to obtain further species information (Fig. 4).

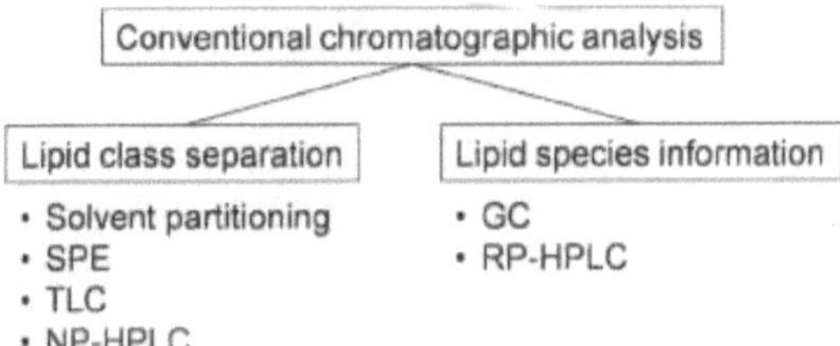

Figure 4. The conventional chromatographic analysis. Two stages in lipid analysis: the separation of lipid classes and the detection of lipid species. SPE: solid phase extraction; TLC: thin-layer chromatography; NP-HPLC: normal-phase high-performance liquid chromatography; GC: gas-liquid chromatography; RP-HPLC: reverse-phase high-performance liquid chromatography.

Several classical strategies can be applied to reduce the sample complexity or separate lipid classes according to their affinity towards different solvents and stationary phases. In this context, methods include solvent partitioning (Bligh and Dyer 1959, Matyash *et al.* 2008), solid phase extraction (SPE) (Ruiz-Gutiérrez and Pérez-Camino 2000); thin-layer chromatography (TLC) (Mangold 1961), and normal-phase high-performance liquid chromatography (NP-HPLC) (Christie 1985, Homan and Pownall 1989) have been commonly used. The major disadvantage of the classical methods is that it usually requires large sample amount and laborious work which consumes considerable time and human resources. Nevertheless, the classical method or a combination of the methods described above can achieve highly purified lipid fractions that are ideal for subsequent absolute quantification.

After lipid separation, the detection approaches such as GC and reverse-phase (RP)-HPLC are proceeded with to obtain further species information. GC analysis is highly robust and suitable for absolute quantification. However, the preliminary criteria of the analytes are that they must be volatile and stable under strong heat. Many compounds therefore require additional derivatization prior to the detection. For instance, derivatization of FSs with *N,O*-bis(trimethylsilyl)trifluoroacetamide (BSTFA) can be performed to convert them into their trimethylsilyl ether derivatives before GC analysis (Diekman and Djerassi 1967). In addition, acidic methanolysis can be applied on lipids (Miquel and Browse 1992), to release their fatty acyl moieties to be analyzed via GC as methyl esters. However, with this approach, the molecular information of individual lipid species (i.e. the fatty acyl moieties) is therefore sacrificed. Nevertheless, GC is still broadly used to quantify purified lipid extracts and obtain first insight into the fatty acid composition of novel biological samples. On the other hand, RP-HPLC can separate lipid species based on the principle of hydrophobic interactions and thus reveal their structural information. It has been widely applied on the analyses of various analytes such as lipids (Moreau 1990), protein variants (Baudin and Wajcman 1987), phenolic compounds (Proestos *et al.* 2005, Proestos *et al.* 2006) and pharmaceuticals (Jeschek *et al.* 2016). In addition, it can be used for specialized targeted lipid analyses such as the fluorescence detection of LCBs after derivatization with *o*-phthalaldehyde (OPA) or 6-aminoquinolyl-*N*-hydroxysuccinimidyl carbamate (AQC) reagents (Abbas *et al.* 1994, Lester and Dickson 2001). In this work, combinations of TLC and GC approaches with precolumn derivatizations according to the analytes have been applied on analyzing the fatty acid content of lipid extracts quantitatively.

1.4.2 Mass spectrometric analysis

The invention and development of mass spectrometers have greatly enhanced the sensitivity and selectivity of chemical analysis. The introduction of soft ionization techniques such as electrospray ionization (ESI) and matrix-assisted laser desorption ionization (MALDI) enlarges its applicability towards biomolecules like proteins and lipids. After ionization, the analytes are separated according to the mass to charge ratio (m/z) by a mass analyzer of choice depending on the analytes and the purpose of the analysis. The commonly used mass analyzers include time-of-flight (TOF), linear quadrupole (Q), linear quadrupole ion trap (LIT), quadrupole ion trap (IT), ion cyclotron resonance (ICR) and orbitrap. In addition, these mass analyzers can be further configured in tandem systems such as triple quadrupole (QqQ), Q-TOF, Q-Orbitrap and so on to perform high-resolution analysis (El-Aneed *et al.* 2009, Holcapek *et al.* 2018).

1.4.2.1 Direct infusion – mass spectrometry

The name DI-MS already suggests its operating principle, which infuses samples directly without prior separation into the MS system. As all the analytes are ionized simultaneously in the identical solvent mixture, DI-MS is suitable for absolute quantification. Therefore, DI-MS has initiated the omics study on lipids, so called shotgun lipidomics (Brügger *et al.* 1997, Han and Gross 1994). The most abundant lipids, mostly glycerophospholipids, in biological extracts have been identified and structurally characterized via shotgun lipidomics including plants (Welti and Wang 2004). However, equal ionization among all analytes in DI-MS analysis is difficult to achieve in a complex biological extract due to the strong matrix effect including the ion suppression and isobaric interference (by compounds which possess the same *m/z* values). This strongly influences the research on less abundant or minor lipid classes, although they may exert critical functions in the cell. Nevertheless, DI-MS with the capability of tandem MS (MS/MS) analysis plays an excellent role in structural characterization of purified compounds. It has been therefore applied on the methodological development of specific biomolecules in this work, in optimizing the parameters of compound ionization and mass analysis.

1.4.2.2 Liquid chromatography – mass spectrometry and its challenges in absolute quantification

To reduce the matrix effect derived from the complex biological extracts, several liquid chromatographic separation approaches have been introduced to the lipidomics setup hereto. The most commonly setup incorporates a reverse-phase (RP)-LC system to the MS, which elutes the compounds according to increasing hydrophobicity, from the column to the subsequent ionizer and the mass spectrometer. Hydrophilic interaction liquid chromatography (HILIC) which separates the analytes according to the hydrophilicity (according to the head groups in case of lipid molecules) has also been introduced to purify the samples and enhance the selectivity of the analysis (Song *et al.* 2018). To analyze very low abundant biomolecules in highly complex samples, an innovative approach that combines HILIC and RP-LC has been developed to perform 2-dimensional LC-MS/MS analysis (Holčapek *et al.* 2015, Li *et al.* 2013, Nie *et al.* 2010, Pham *et al.* 2019).

The lipid analysis in this work majorly relies on an RP-LC-MS/MS system equipped with a QqQ, which provides the molecular information of individual lipid species with the multiple reaction monitoring (MRM) detection mode (Fig. 5). With the MRM mode, precursor molecules are selected in the first mass analyzer (Q1) and fragmented in the second mass analyzer (Q2); specific fragments are then analyzed in the third mass analyzer (Q3). With this strategy, a highly selective analysis detecting specific pairs of precursor – fragment can be applied on complex biological extracts, which enables the development of the targeted lipidomics platform. Previous work by Dr. Pablo Tarazona has achieved to establish a wide-ranging lipidomics platform targeting more than 300 lipid species from *Arabidopsis* leaves (Tarazona *et al.* 2015). However, this platform lacks a few lipid classes that exert critical biological functions due to their low abundance. To broaden and enhance the pre-existing lipidomics workflow is thus necessary for a comprehensive understanding of the biological membranes.

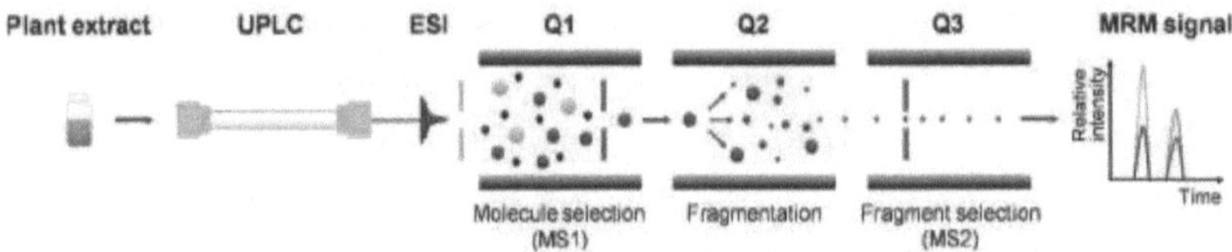

Figure 5. The targeted lipidomics workflow in this work. The lipid molecules are separated by UPLC and ionized by soft ionization before analyzing by MS/MS with MRM mode. UPLC: ultra-performance liquid chromatography; ESI: electrospray ionization: MRM: multiple reaction monitoring.

Even though liquid chromatography, which separates the analytes and elutes them at distinct time points has contributed to the great enhancement of the sensitivity and selectivity in LC-MS analysis, it creates a major obstacle in the absolute quantification. Although reliable relative amounts of the analytes can be revealed via LC-MS based analysis (Cajka and Fiehn 2014), many factors hinder the quantification of their absolute abundance. First, along the liquid chromatographic gradient, different analytes can have unequal solubility at different time points according to the distinct solvent mixture. That is, higher abundant analytes with low solubility in the solvent mixture may arrive in lower numbers at the detector compared to lower abundant analytes with higher solubility (Snyder *et al.* 2010). Second, the simultaneously eluted analytes can interact with homo- and hetero-molecules in the ionization source and compete for ionization efficiency. This results in reduced detector response and / or decreased signal-to-noise ratio, so called ion suppression (Annesley 2003, Taylor 2005). Moreover, the alterations of the solvent mixture along the chromatographic gradient leads to instable ionization and can result in different detector responses (Kostiainen and Kauppila 2009). Finally, the kinetics of the analyte-specific fragments depends on the size, structure and charge of the distinct fragments generated in the collision cell of the mass spectrometer. Therefore, different fragments of a specific precursor analyte can contribute to uneven signal intensities from the detector as well (Demarque *et al.* 2016).

Several studies have been pursued to overcome these restrictions. The best practice is to create a standard curve with the targeted analytes in correlation to the internal standard that is eluted at the same retention time along the chromatographic gradient. In addition, the sample matrix used for generating the standard curves, the surrogate matrix, should possess a comparable composition as the real samples (Krautbauer *et al.* 2016, Wakamatsu *et al.* 2018). However, it is difficult to achieve these criteria including finding the identical internal standards for the wide-ranging analytes and a matching matrix for a high-throughput omics analysis for biological samples. In case of lipidomics analysis, several endogenous lipid components (such as Cer-Ps and GIPCs) are commercially unavailable. Therefore, standard curves are normally prepared using a few selected lipid species of certain lipid classes and / or purchasable structural analogues, although it has been demonstrated that the signals obtained from the simulated compounds are statistically different to the endogenous analytes (Dahal *et al.* 2011). A recent study has applied a regression algorithm on predicting the responses of the analytes in non-targeted LC-MS screening (Liigand *et al.* 2020). Nevertheless, whether the model provides reliable accuracy for absolute quantification in lipidomics analysis requires cautious verification. In an attempt to obtain a comprehensive and quantitative lipid composition of the plant PM, a combinatorial approach that incorporates the qualitative lipid species-targeting LC-MS and the quantitative lipid class-targeting TLC-GC analyses was conducted in this study. Specific ratios between lipid species within each lipid class can be thereby revealed and thus contribute to depict the lipid landscape of the plant PM.

1.5 Aims of this study

The maintenance of the structural and functional integrities of the biological membranes is critical for the survival of cells and their organelles. The overall composition of cellular lipids has been studied intensively in the past centuries. However, our knowledge concerning the functions of individual membrane lipids and the mechanisms of their compositional remodeling at the subcellular level is still in its infancy. The main objective of this study was to enhance and broaden the pre-existing lipidomics workflow by incorporating the functional but minor lipid classes including phosphoinositides, complex glycosphingolipids and phosphorylated sphingolipids (Chapter 2 and 3) and to apply this method to selected membrane preparations from designated Arabidopsis plants to ultimately reveal their (species specific) functions (Chapter 4 and 5).

The plasma membrane (PM) is the exterior border of the cell. Understanding its compositional remodeling under stressed conditions is of high agronomic and economic interest. The establishment of the ordered lipid microdomains, the lipid rafts, within the PM is mediated amongst others by α-hydroxylated sphingolipids. In this respect, detailed lipidomics analyses were conducted to profile the cold-induced alterations between the compositions of the purified PMs from *Arabidopsis* wild type and the sphingolipid α-hydroxylase mutant *fah1 fah2*. Furthermore, the role of the α-hydroxylated sphingolipids on the PM organization under both normal and CA was investigated. Noteworthy, the transversal distribution of individual lipid species across the PM was also evaluated by this advanced lipidomics method in combination with the purification of differently orientated PM vesicles, to assess the functional roles they may exert within the distinct membrane leaflet (Chapter 4).

The knowledge concerning the detailed composition and the functions of lipids within mitochondria is still scarce. In addition, although the mitochondrion can synthesize parts of its lipids on its own, it highly relies on the import of lipid moieties from other organelles, especially through membrane contact sites as it is disassociated from the vesicular transport. To elucidate its capacity of lipid biosynthesis and modification, a combinatorial approach integrating the in-depth lipidomics with proteomics and online database mining was conducted to profile the mitochondrial lipidome including glycerolipids, sphingolipids and sterols, to identify membrane contact site-localized proteins and to construct possible lipid biosynthesis pathways in plant mitochondria (Chapter 5).

Chapter 2.

Targeted analysis of the plant lipidome by UPLC-nanoESI-MS/MS

The article was submitted in March 2020 as a chapter in the book *Molecular Biology Plant Lipids: Plant Lipids, Methods and Protocolls*.

Author contribution

Yi-Tse Liu performed the methodological developments to include and / or enhance the analyses of phosphatidyl mono- and bisphosphates, cardiolipins, sulfoquinovosyl diacylglycerols, long-chain base phosphates, ceramide phosphates, series A and B glycosyl inositol phosphoceramides in the pre-existing lipidomics workflow. The conducted experiments were based on fractionation of subcellular membrane, chemical derivatization of the phosphorylated lipids and optimization of the LC-MS/MS parameters fine-tuning by DI-MS.

Targeted Analysis of the Plant Lipidome by UPLC-NanoESI-MS/MS

Cornelia Herrfurth[1, 2], Yi-Tse Liu[1] and Ivo Feussner[1, 2, 3, *]

[1]University of Goettingen, Albrecht-von-Haller-Institute for Plant Sciences, Department of Plant Biochemistry, 37077 Goettingen, Germany

[2]University of Goettingen, Goettingen Center for Molecular Biosciences (GZMB), Service Unit for Metabolomics and Lipidomics, 37077 Goettingen, Germany

[3]University of Goettingen, Goettingen Center for Molecular Biosciences (GZMB), Department of Plant Biochemistry, 37077 Goettingen, Germany

*Correspondence: Ivo Feussner, University of Goettingen, Albrecht-von-Haller-Institute for Plant Sciences, Department of Plant Biochemistry, Justus-von-Liebig-Weg 11, 37077 Goettingen, Germany; Fax:+49-551-39-5749; Email: ifeussn@uni-goettingen.de

Running head: Plant Lipidomics Analysis

Key words Glycerolipids, Sphingolipids, Sterol lipids, Liquid chromatography, Nanoelectrospray ionization, Mass spectrometry, Membrane enrichment, Monophasic extraction, Methylation, Acetylation

1 Introduction

Lipids can function as energy source, structural components of cellular membranes, and signaling molecules (Shulaev and Chapman 2017). Thousands of lipid molecular species exist in different organisms in a concentration range spanning at least six orders of magnitude, and these lipids can be categorized based on their molecular structure (Fahy *et al.* 2011, Liebisch *et al.* 2013). Lipidomics is an analytical discipline to study lipid metabolism on a broad scale with mass spectrometric techniques (Han and Gross 2003). Due to the development of mass spectrometry (MS) over the last twenty years, the lipidomic technology has greatly advanced with respect to its analytical range, detection sensitivity, and speed of analysis. Lipidomic studies aim to detect the complete set of lipids (the lipidome) of a given cell type or tissue for understanding their relation and function within the context of cellular metabolism (Yang and Han 2016).

Two major analytical approaches exist, namely, global and targeted analyses. These can be distinguished by their analytical coverage and lipidomic applications, as well as by the presence or absence of liquid chromatography (LC) prior to the MS analysis. For global or "shotgun lipidomics", a lipid extract is directly infused into the MS system and analyzed without prior chromatographic separation (Samarakoon *et al.* 2012, Simons *et al.* 2012). The entire lipidome is recorded at a constant sample concentration, and the identification and quantification of the molecular lipid species is performed without limitation on the acquisition time other than sample volume. The complexity of the sample matrix, however, limits the selectivity for isomeric and isobaric lipid species, as well as the detection sensitivity for trace lipid species (e.g. those with signaling functions). This restriction can be overcome by the use of LC separation prior to tandem MS (MS/MS) analyses. The targeted methodology reported herein has been developed to perform a sensitive and highly resolved analysis, screening molecular species of a total of 36 lipid subclasses from a minimal amount of plant material. The workflow includes a monophasic propan-2-ol/hexane/water extraction. It was originally developed for the efficient extraction of amphiphilic sphingolipids (Markham *et al.* 2006) and is applied here to extract a broad range of plant lipids with highly diverse chemical properties. Prior to the MS-based lipidomics analysis, the overall lipid building blocks, namely acyl chains, polar head groups, and backbones, are determined by classical lipid analytical methods (e.g. GC, TLC). This information is then used to calculate an array including all possible combinations of putative plant molecular lipid species. Based on this array, the putative precursor ions and corresponding lipid subclass-specific fragment ions (Table 2) are derived and converted into mass transition lists for the MS/MS detection. The lipid extract is subjected to an ultra-performance LC (UPLC) system coupled with a chip-based nanoelectrospray ionization (nanoESI) source and a triple quadrupole analyzer. The robust and efficient resolving power of the sequential UPLC separation and targeted MS/MS detection enables the analysis of the distinct acyl composition of the molecular species within most lipid subclasses. For triacylglycerols, however, only the averaged composition corresponding to the total number of carbon atoms and double bonds can be resolved due to the wide range of possibilities for its acyl combinations in a single targeted molecule. The specificity of the UPLC-nano ESI-MS/MS method is additionally increased by incorporating chemical derivatization approaches (methylation and acetylation) after lipid extraction. Thus, distinct functional lipid groups, such as trace phospholipid species, can be detected in plant tissues.

2 Materials

2.1 Samples and Buffers

1. Flash-frozen (i.e. in liquid nitrogen) plant tissue (various developmental stages, e.g. seeds, seedlings, leaves), keep at- 80 °C until grinding and extraction.

2. Cell cultures from plants, algae and yeast, keep at-80 °C until extraction.

3. Microsomal fractions, keep at - 80 °C under argon until extraction.

4. Microsome Extraction Buffer (MEB) 1: 0.1 M Tris-HCl, pH 7.5, 0.81 M sucrose, 5 % (*v/v*) glycerol, 10 mM ethylenediaminetetraacetic acid (EDTA), pH 8.0, 10 mM ethyleneglycoltetraacetic acid

(EGTA), pH 8.0, 5 mM KCl, 1 mM 1,4-dithiothreitol (DTT), 1 mM phenylmethanesulfonyl fluoride (PMSF).

5. Lipid Extraction Buffer (LEB): propan-2-ol/hexane/water (60:26:14, *v/v/v*)

2.2 Chemicals and Standards

Analytical standards were purchased from Merck KGaA (Darmstadt, Germany), Avanti Polar Lipids, Inc. (Alabama, AL, USA) and Matreya (State College, PA, USA).

1. Trimethylsilyldiazomethane solution for methylation: 2 M in hexane (Merck KGaA, Darmstadt, Germany).

2. Pyridine, acetic anhydride for acetylation.

3. Tetrahydrofuran/methanol/water (4:4:1, *v/v/v*) for dissolving lipid samples.

4. Methylamine for glycerolipid hydrolysis: 33 % (*v/v*) methylamine in ethanol.

5. *N,O*-bis(trimethylsilyl)trifluoroacetamide (BSTFA), pure for silylation (Merck KGaA, Darmstadt, Germany).

2.3 Solvents and Solutions for LC-MS

All solvents used (methanol, propan-2-ol, tetrahydrofuran) are LC-MS grade quality unless indicated otherwise. Ultra-pure water is always freshly generated by an Arium pro VF TOC ultrapure water system (Sartorius, Goettingen, Germany).

1. Solvent system for UPLC analyses with the ACQUITYHSS T3 column (Waters Corporation, Milford, MA, USA): solvent A (methanol/20 mM ammonium acetate, 3:7, *v/v*, containing 0.1 %, *v/v* acetic acid), solvent B (tetrahydrofuran/methanol/20 mM ammonium acetate, 6:3:1, *v/v/v*, containing 0.1 %, *v/v* acetic acid).

2. Tuning mixture for QTRAP6500: Standards chemical kit with low/high concentration polypropylene glycols (PPGs) (AB Sciex, Framingham, MA, USA).

2.4 LC-MS System

1. For chromatographic separation: ACQUITY UPLC® system (Waters Corporation, Milford, MA, USA) equipped with an ACQUITY UPLC® HSS T3 column (100 mm x 1 mm, 1.8 µm; Waters Corporation, Milford, MA, USA). This is a silica-based, reversed-phase C18 column.

2. For nano ESI: chip ion source TriVersa Nanomate® (Advion, Incorporation, Ithaca, NY, USA) equipped with nanoESI chip with 5 µm internal diameter nozzles.

3. For mass-spectrometric detection: AB Sciex QTRAP6500® tandem mass spectrometer (AB Sciex, Framingham, MA, USA).

2.5 Software

1. Data acquisition: Analyst 1.6.2 (AB Sciex, Framingham, MA, USA).

2. nanoESI control: ChipSoft 8.3.1 (Advion, Incorporation, Ithaca, NY, USA).

3. Data analysis: MultiQuant 3.0.2 (AB Sciex, Framingham, MA, USA).

4. Data processing and statistics: Excel 2016 (Microsoft Corporation, Redmond, WA, USA) and RStudio (RStudio, Incorporation, Boston, MA, USA).

2.6 Other Equipment

1. Kimble extraction tubes (Kimax-51, 13 x 100 mm) with Teflon-lined screw caps (Gerresheimer Glass Inc., Vineland, NY, USA).

2. Chemically resistant tips for organic solvents (Safe Seal Tips Premium, Biozym, Oldendorf, Germany).

3. Test strips for semi-quantitative determination of hydrogen peroxide and peroxides in tetrahydrofuran (Quantofix® Peroxide 100, Macherey-Nagel, Dueren, Germany).

4. Glass sample vials for sample storage (1.1 ml, inner cone in the solid glass bottom, ND9, VWR International GmbH, Darmstadt).

5. Glass micro vials for analysis (12 mm, 250 µl, Macherey-Nagel GmbH, Dueren, Germany) fixed by a spring in HPLC glass vials (1.5 ml, N9, Macherey-Nagel GmbH, Düren, Germany).

6. Nitrogen evaporator (Organomation Associates, Incorporation, Berlin, MA, USA).

7. Mixer Ball Mill MM200 with stainless steel grinding jars or PTFE-jars (Retsch GmbH, Haan, Germany).

8. Freeze dryer.

9. Argon.

3 Methods

3.1 Harvesting and Homogenization of Plant Material

1. Complete the harvesting of plant material as quickly as possible and always in the same time range to avoid unspecific lipid degradation by lipolytic activities.

2. Shock freeze the harvested material immediately in liquid nitrogen (*see* **Note 1**).

3. Grind the plant material by using a porcelain mortar and pestle or homogenize using the Mixer Ball Mill MM200 (*see* **Note 2**).

4. Use stainless steel grinding jars (for large sample amounts) or PTFE-jars for Eppendorf tubes (for small sample amounts) with the corresponding size of stainless steel balls for mill homogenization.

5. Time and repetitions of the homogenization cycles depends on the amount and rigidity of the biological material (*see* **Note 3**).

6. Ensure that the biological material always stays completely frozen under liquid nitrogen.

7. For freeze drying, incubate the sample material in the freeze dryer overnight until the pressure stays constant, indicating that the water of the sample has been completely removed.

3.2 Enrichment of Microsomal Membrane Fractions

To concentrate minor lipids located in cellular membranes, microsomal-type membranes are enriched from small amounts of plant material. Microsomal membranes are isolated without the need for ultracentrifugation (modified from (Abas and Luschnig 2010)):

1. Prior to use, the 2 ml Eppendorf tubes are kept on ice. All solvents are kept at room temperature.

2. Weigh 50 mg of homogenized deep frozen material, or 5 mg of homogenized freeze-dried material, into a 2 ml Eppendorf tube (*see* **Note 4**). Immediately add 0.2 ml of the extraction buffer (MEB) 1. Ensure that the biological material is completely covered with the MEB 1.

3. Vortex the sample strongly.

4. Centrifuge the samples for 3 min at 600 *g* and 4 °C.

5. Transfer the supernatant into a new 2 ml Eppendorf tube and put it aside on ice. Re-extract the sample with 0.1 ml MEB 2 (dilute MEB 1 to 0.35x in water: mix 35 ml of MEB 2 with 65 ml of water).

6. Vortex the sample strongly.

7. Centrifuge the samples for 3 min at 600 *g* and 4 °C.

8. Add the supernatant of this second extraction to the first supernatant. Re-extract the sample with 65 µl MEB 3 (dilute MEB 1 to 0.48x in water: mix 48 ml of MEB1 with 52 ml of water).

9. Vortex the sample strongly.

10. Centrifuge the samples for 30 s at 2,000 *g* and 4 °C.

11. Transfer the supernatant to the combined supernatants. Add 0.25 ml water.

12. Vortex the pooled supernatant strongly. Transfer three 0.2 ml aliquots into 2 ml Eppendorf tubes.

13. Centrifuge the samples for 2 h at 20,000 *g* and 4 °C to obtain the membrane pellets.

14. Remove the supernatant carefully. Wash the membrane pellet with 0.15 ml water.

15. Centrifuge the samples for 45 min at 20,000 *g* and 4 °C.

16. Remove the supernatant carefully.

17. Cover the membrane pellet with argon and use it either immediately for monophasic lipid extraction or store it at -20 °C until extraction.

To avoid autoxidation of lipids, immediately cover the samples with argon after each extraction step, particularly at the end of membrane isolation procedure before storage.

3.3 Extraction of Lipids from Plant Material and Cultured Cells

The monophasic extraction method with propan-2-ol, hexane, and water as described by Markham *et al.* (Markham *et al.* 2006)was slightly modified as follows (Grillitsch *et al.* 2014):

1. Prior to use, Kimble glass tubes and the lipid extraction buffer (LEB) (propan-2-ol/hexane/water (60:26:14, *v/v/v*)) are warmed to 60 °C.

2. Weigh 200 mg of homogenized and deep frozen material, or 20 mg of homogenized and freeze-dried material, into Kimble glass tubes (*see* **Note 4**). Immediately add 6 ml of the warmed LEB (*see* **Note 5**). Ensure that the biological material is completely covered with the extraction buffer.

3. Vortex the sample strongly.

4. Shake for 30 min at 60 °C. During this incubation process, vortex and sonicate the sample every 10 min.

5. Centrifuge the samples for 20 min at 800 *g* and 20 °C.

6. Collect the supernatant with a glass Pasteur pipette and transfer it into a new Kimble glass tube.

7. Dry the supernatant under a nitrogen stream.

8. Dissolve the samples in 0.8 ml of tetrahydrofuran/methanol/water (4:4:1, *v/v/v*). Ensure that no material is stuck to the wall of the glass tube.

9. Centrifuge (5 min, 800 x g, 20 °C) and transfer the supernatant into a glass sample vial.

10. Cover the sample with argon and store it at -20 °C until UPLC-nanoESI-MS/MS analysis (*see* **Note 6**).

11. Transfer aliquots (30 - 100 μl) of the sample into glass micro vials directly before starting the analysis.

To avoid autoxidation, immediately cover the samples with argon after each extraction step, particularly at the end of the extraction procedure before storage.

3.4 Extraction of Lipids from Microsomal Membrane Fractions

The monophasic extraction method with propan-2-ol, hexane, and water as described by Markham *et al.* (Markham *et al.* 2006) was modified for the extraction of microsomal membrane fractions as follows:

1. Prior to use, Kimble glass tubes and the LEB are warmed to 60 °C. Also warm propan-2-ol and hexane, which are added separately to the sample after resuspension of the pellets in water.
2. Combine the three microsomal membrane pellets derived from a single sample in 0.14 ml water and transfer the sample into a Kimble glass tube (*see* **Note 4**). Immediately add the warmed 0.6 ml propan-2-ol and the 0.26 ml hexane (*see* **Note 5**), and 5 ml of warmed LEB.
3. Vortex the sample strongly.
4. Shake for 30 min at 60 °C. During this incubation vortex and sonicate the sample every 10 min.
5. Centrifuge the samples for 20 min at 800 *g* and 20 °C.
6. Collect the supernatant with a glass Pasteur pipette and transfer it into a new Kimble glass tube.
7. Dry the supernatant under a nitrogen stream.
8. Dissolve the samples in 0.2 ml of tetrahydrofuran/methanol/water (4:4:1, *v/v/v*). Ensure that no material is stuck to the wall of the glass tube.
9. Centrifuge (5 min, 800 *g*, 20 °C) and transfer the supernatant into a glass sample vial.
10. Cover the sample with argon and store it at -20 °C until UPLC-nanoESI-MS/MS analysis (*see* **Note 6**) or before chemical derivatization.
11. Transfer aliquots (30 - 100 µl) of the sample into glass micro vials directly before the analysis.

To avoid autoxidation, immediately cover the samples with argon after each extraction step, particularly at the end of the extraction procedure before storage.

3.5 Chemical Derivatization of Lipids

Phosphate groups and hydroxyl groups of some lipid classes are chemically modified by methylation (lysophosphatidic acid, phosphatidic acid, phosphatidylinositol phosphate and phosphatidylinositol bisphosphate) or acetylation (ceramide phosphate, free sterol and long-chain base phosphate) to improve their chromatographic separation and mass spectrometric detection.

1. Transfer an aliquot (30 - 100 µl) of the lipid sample into a glass sample vial (*see* **Note 4**).
2. Dry the aliquot under a nitrogen stream.
3. For methylation (modified from (Lee *et al.* 2013)): dissolve the dry lipid in 0.4 ml methanol and add 6.5 µl of trimethylsilyldiazomethane solution (2 M in hexane). Vortex the sample strongly. Incubate for 30 min at room temperature and terminate the reaction by neutralizing with 2 µl of 1 N acetic acid.
4. For acetylation (modified from (Berdyshev *et al.* 2005)): dissolve the dry lipid in 100 µl pyridine and 50 µl of acetic anhydride. Vortex the sample strongly. Incubate for 30 min at 50 °C.
5. Dry the derivatized sample aliquot under a nitrogen stream.
6. Dissolve the sample aliquot in an equal volume of tetrahydrofuran/methanol/water (4:4:1, *v/v/v*) as before the chemical derivatization. Ensure that no material is stuck to the wall of the glass tube.
7. Centrifuge (5 min, 800 *g*, 20 °C) and transfer the sample into a glass micro vial.
8. Cover the derivatized sample aliquot with argon and store it at -20 °C until UPLC-nanoESI-MS/MS analysis (*see* **Note 6**).

To avoid autoxidation, immediately cover the sample aliquot with argon after each derivatization step, particularly at the end of the derivatization procedure before storage.

3.6 Methylamine Treatment for Enhanced Sphingolipid Analysis

To improve the detection efficiency for sphingolipids, the lipid extract is treated with methylamine. This mild base hydrolyzes glycerophospholipids, but not sphingolipids, and therefore reduces interference specifically for sphingolipid analysis, for example during the ESI process (Markham and Jaworski 2007).

1. Transfer 200 µl of the extracted lipid into a glass sample vial.
2. Dry the aliquot under a nitrogen stream.
3. Add 1.4 ml of 33 % (*v/v*) methylamine in ethanol and 0.6 ml of water (modified from (Markham 2013)).
4. Vortex the sample strongly. Incubate for 1 h at 50 °C.
5. Dry the methylamine-treated lipid under a nitrogen stream.
6. Dissolve the methylamine-treated lipid in 50 µl of tetrahydrofuran/methanol/water (4:4:1, *v/v/v*). Ensure that no material is stuck to the wall of the glass tube.

7. Centrifuge (5 min, 800 x g, 20 °C) and transfer the supernatant into a glass micro vial.

8. Cover the methylamine-treated lipid aliquot with argon and store it at - 20 °C until UPLC-nanoESI-MS/MS analysis (*see* **Note 6**).

3.7 Lipid Analysis by UPLC-NanoESI-Mass Spectrometry

1. Set the temperature of the autosampler (sample manager) of the UPLC system to 18 °C and the column oven temperature to 40 °C.

2. Set the flow rate to 0.1 ml/min or 0.13 ml/min (depending on the gradient; Table 1) and the injection volume to 2 µl.

Table 1 Solvent gradients for UPLC separation (*see* **Note 6**) prior to detection by mass spectrometry.

Gradient	Time (min)	Flow (ml/min)	Solvent A (%)	Solvent B (%)
1	0	0.13	10	90
	2	0.13	10	90
	4	0.13	0	100
	8	0.13	0	100
	8.5	0.13	10	90
	12	0.13	10	90
2a	0	0.1	20	80
	2	0.1	20	80
	10	0.1	0	100
	12	0.1	0	100
	12.5	0.1	20	80
	16	0.1	20	80
2b	0	0.1	35	65
	2	0.1	35	65
	10	0.1	0	100
	12	0.1	0	100
	12.5	0.1	35	65
	16	0.1	35	65
2c	0	0.1	60	40
	2	0.1	60	40
	10	0.1	0	100
	12	0.1	0	100
	12.5	0.1	60	40
	16	0.1	60	40

3. Use methanol as strong and methanol/water (1:9, *v/v*) as weak wash solutions.

4. Use the gradients of solvents shown in Table 1 as mobile phase for the chromatographic separation, depending on the lipid classes of interest (Table 2) (Grillitsch *et al.* 2014). The retention time windows for the elution of the corresponding lipid subclasses are shown in

Figure 1. Representative chromatograms illustrating the separation by the different gradients have been published in (Tarazona *et al.* 2015).

5. The performance of the UPLC should be controlled regularly by inspecting the back pressure of the system and the retention time stability using lipid extracts with known composition or analytical standards.

6. Set the ionization voltage of the NanoESI system to -1.5 kV in negative mode, or to 1.5 kV in positive mode, when the UPLC flow is started.

7. The performance of the nanoESI device has to be controlled regularly by visually inspecting the surface of the chip and routinely calibrating the LC coupler using ChipSoft. During the analysis, the nanoelectrospray current should be monitored constantly.

8. Operate the QTRAP6500® tandem mass spectrometer in MRM mode in either negative or positive mode depending on the lipid class of interest (Table 2).

9. Import lipid subclass-specific mass transition lists of molecular species constructed on the basis of the identified lipid building blocks (see Section 3.11). The general calculation of the precursor *m/z* values, product *m/z* values, and the optimized MS parameters are shown in Table 1.

10. Set the dwell time to 5 msec for all mass transitions.

11. Adjust the resolution of the mass analyzers to 0.7 amu full width at half-height (FWHH).

12. Set the ion source temperature to 40 °C and the curtain gas at 10 (given in arbitrary units).

13. The performance of the QTRAP6500® tandem mass spectrometer has to be controlled regularly. Inspection of the mass sensitivity with lipid extracts with known composition or analytical standards should be performed before running samples. The mass spectrometer has to be cleaned annually following the instruction of the manufacturer. Moreover, the mass accuracy and resolution have to be tuned using the tuning mixture following cleaning.

14. For identification of precursor ions and fragment ions, use the Q1 MS mode or the product ion mode and vary the declustering potential and collision energy depending on the requirements of the analyte ion.

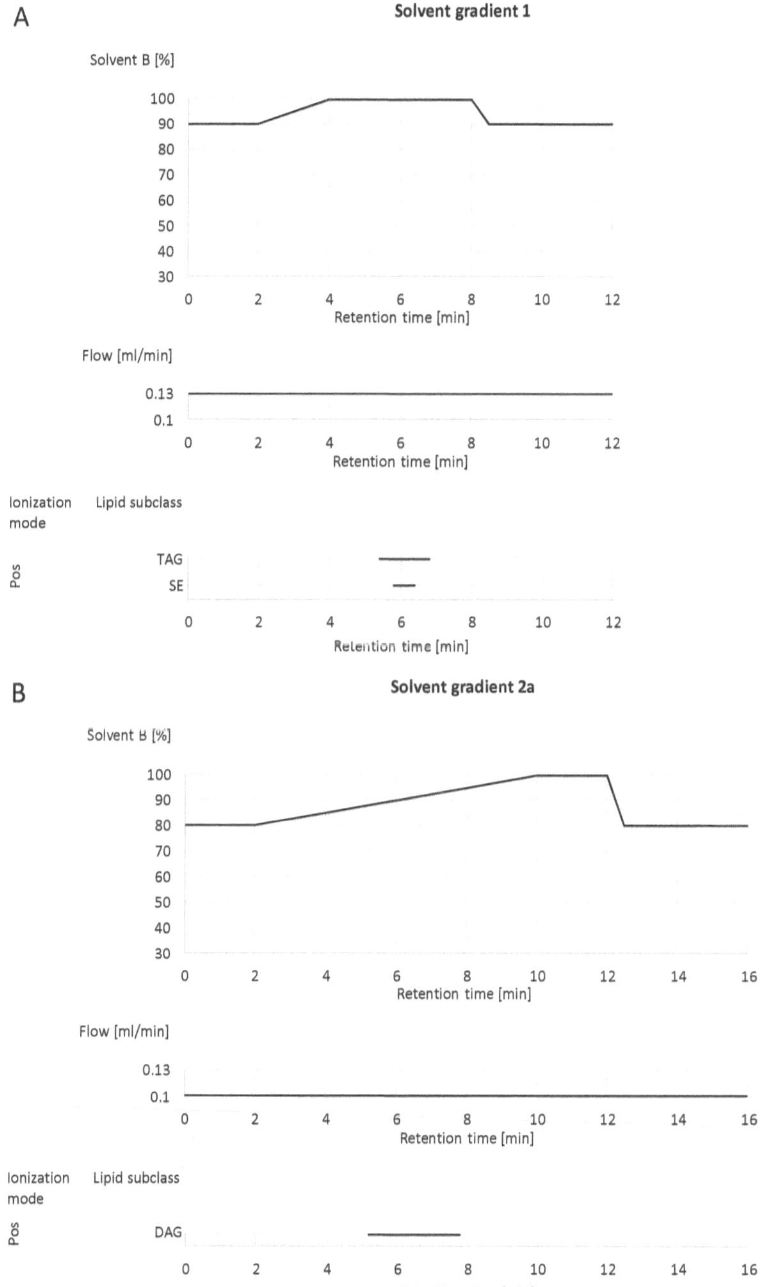
A
Solvent gradient 1
Solvent B [%]
100
90
80
70
60
50
40
30
0 2 4 6 8 10 12
Retention time [min]
Flow [ml/min]
0.13
0.1
0 2 4 6 8 10 12
Retention time [min]
Ionization mode
Lipid subclass
Pos
TAG
SE
0 2 4 6 8 10 12
Retention time [min]
B
Solvent gradient 2a
Solvent B [%]
100
90
80
70
60
50
40
30
0 2 4 6 8 10 12 14 16
Retention time [min]
Flow [ml/min]
0.13
0.1
0 2 4 6 8 10 12 14 16
Retention time [min]
Ionization mode
Lipid subclass
Pos
DAG
0 2 4 6 8 10 12 14 16
Retention time [min]

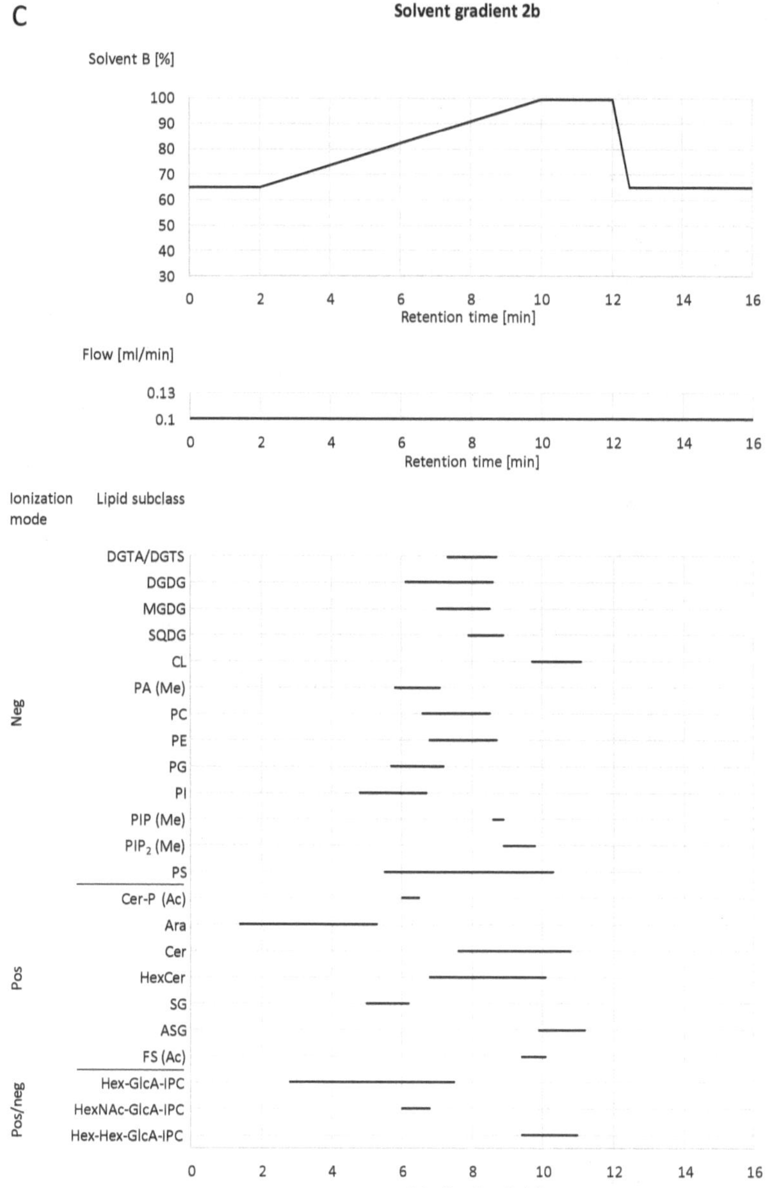

C
Solvent gradient 2b
Solvent B [%]
100
90
80
70
60
50
40
30
0 2 4 6 8 10 12 14 16
Retention time [min]
Flow [ml/min]
0.13
0.1
0 2 4 6 8 10 12 14 16
Retention time [min]
Ionization mode
Lipid subclass
Neg
DGTA/DGTS
DGDG
MGDG
SQDG
CL
PA (Me)
PC
PE
PG
PI
PIP (Me)
PIP2 (Me)
PS
Pos
Cer-P (Ac)
Ara
Cer
HexCer
SG
ASG
FS (Ac)
Pos/neg
Hex-GlcA-IPC
HexNAc-GlcA-IPC
Hex-Hex-GlcA-IPC
0 2 4 6 8 10 12 14 16
Retention time [min]

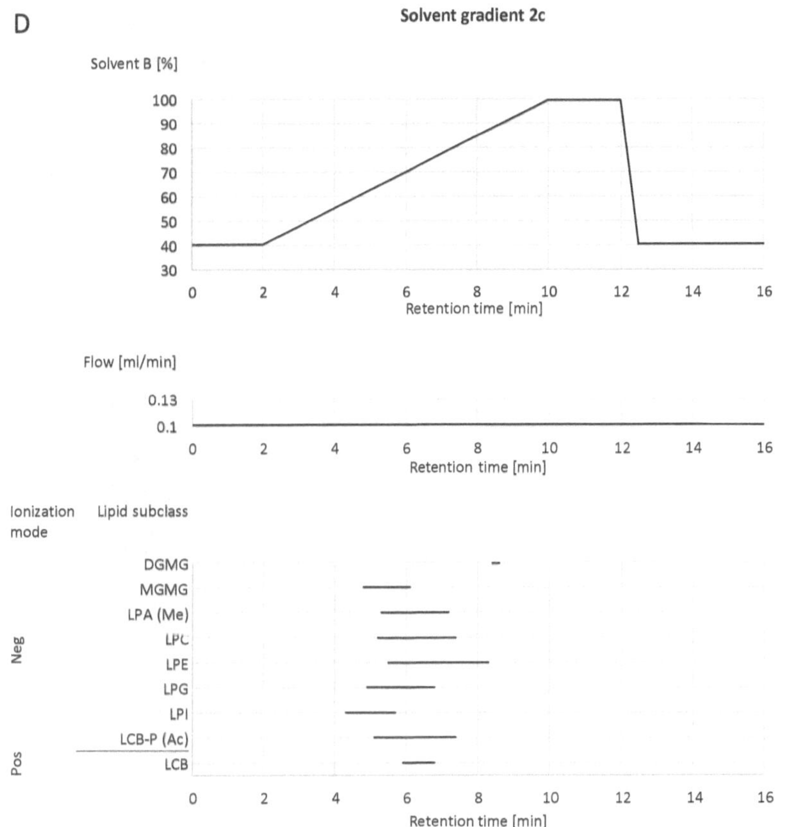

Figure 1 Solvent gradient-depending retention time windows for the elution of the corresponding lipid subclasses by the UPLC-nanoESI-MS/MS analysis.

The amount of solvent B, the flow rate, the ionization mode and the retention time window for the elution of lipids are shown for solvent gradient 1 (A), solvent gradient 2a (B), solvent gradient 2b (C) and solvent gradient 2c (D). All data derive from the UPLC-nanoESI-MS/MS analysis of lipid extracts from leaves of *Arabidopsis thaliana* except Cer-P, HexNAc-GlcA-IPC, Hex-Hex-Glc-IPC, LCB-P (microsomal membrane fractions (Zienkiewicz *et al.* 2020)), CL (isolated mitochondria), DGTA (*Physcomitrella patens*). Abbreviations are explained in Table 2.

Table 2 Mass transitions and optimized MS parameters for detection of molecular species from various lipid classes by mass spectrometry.

Lipid Category Class	Subclass	Solvent gradient	Ionization mode	Q1	Q3	DP [V]	EP [V]	CE [V]	CXP [V]	Reference
Glycerolipids										
Betaine lipids	DGTA/DGTS	2b	Negative	[M+OAc]$^-$	[RCOO]$^-$	-180	-10	-60	-13	[f]
	DGCC	2b	Negative	Negative	[RCOO]$^-$	-180	-10	-60	-13	[f]
Glycolipids	Ara Ara-A, Ara-B Ara-C, Ara-D Ara-E, Ara-G	2b	Positive	[M+NH$_4$]$^+$	[M-(monoGal-H)+NH$_4$]$^{+\ c}$ [M-(diGal-H)+NH$_4$]$^{+\ d}$ [M-(monoGal$_{acylated}$-H)+NH$_4$]$^{+\ e}$	100	10	30	6	(Ibrahim *et al.* 2011)
	DGDG	2b	Negative	[M+OAc]$^-$	[RCOO$_{sn1}$]$^-$ / [RCOO$_{sn2}$]$^-$	-100	-10	-40	-10	(Tarazona *et al.* 2015)
	DGMG	2c	Negative	[M+OAc]$^-$	[RCOO]$^-$	-200	-10	-40	-11	(Tarazona *et al.* 2015)
	MGDG	2b	Negative	[M+OAc]$^-$	[RCOO$_{sn1}$]$^-$ / [RCOO$_{sn2}$]$^-$	-100	-10	-45	-10	(Tarazona *et al.* 2015)
	MGMG	2c	Negative	[M+OAc]$^-$	[RCOO]$^-$	-200	-10	-40	-11	(Tarazona *et al.* 2015)
	SQDG	2b	Negative	[M-H]$^-$	[RCOO$_{sn1}$]$^-$ / [RCOO$_{sn2}$]$^-$	-100	-10	-40	-10	(Tarazona *et al.* 2015)
	SQMG	2c	Negative	[M-H]$^-$	[RCOO]$^-$	-200	-10	-40	-11	(Tarazona *et al.* 2015)
Neutral lipids	DAG	2a	Positive	[M+NH$_4$]$^+$	[M-RCOO]$^+$	100	10	38	10	(Tarazona *et al.* 2015)
	TAG	1	Positive	[M+NH$_4$]$^+$	[M-RCOO]$^+$	140	10	40	6	(Tarazona *et al.* 2015)
Phospholipids	CL	2b	Negative	[M-2H]$^{2-}$	[RCOO]$^-$	-100	-10	-40	-10	(Zhou *et al.* 2016a)
	LPA	2c	Negative[a]	[M+Me-H]$^-$	[RCOO]$^-$	-200	-10	-30	-11	(Lee *et al.* 2013)
	LPC	2c	Negative	[M+OAc]$^-$	[RCOO]$^-$	-200	-10	-40	-11	(Tarazona *et al.* 2015)
	LPE	2c	Negative	[M-H]$^-$	[RCOO]$^-$	-200	-10	-40	-11	(Tarazona *et al.* 2015)
	LPG	2c	Negative	[M-H]$^-$	[RCOO]$^-$	-200	-10	-40	-11	(Tarazona *et al.* 2015)
	LPI	2c	Negative	[M-H]$^-$	[RCOO]$^-$	-200	-10	-40	-11	(Tarazona *et al.* 2015)
	LPS	2c	Negative	[M-H]$^-$	[RCOO]$^-$	-200	-10	-40	-11	(Tarazona *et al.* 2015)
	PA	2b	Negative[a]	[M+Me-H]$^-$	[RCOO$_{sn1}$]$^-$ / [RCOO$_{sn2}$]$^-$	-200	-10	-38	-11	(Lee *et al.* 2013)
	PC	2b	Negative	[M+OAc]$^-$	[RCOO$_{sn1}$]$^-$ / [RCOO$_{sn2}$]$^-$	-100	-10	-40	-10	(Tarazona *et al.* 2015)
	PE	2b	Negative	[M-H]$^-$	[RCOO$_{sn1}$]$^-$ / [RCOO$_{sn2}$]$^-$	-100	-10	-40	-10	(Tarazona *et al.* 2015)
	PG	2b	Negative	[M-H]$^-$	[RCOO$_{sn1}$]$^-$ / [RCOO$_{sn2}$]$^-$	-100	-10	-40	-10	(Tarazona *et al.* 2015)
	PI	2b	Negative	[M-H]$^-$	[RCOO$_{sn1}$]$^-$ / [RCOO$_{sn2}$]$^-$	-100	-10	-40	-10	(Tarazona *et al.* 2015)
	PIP	2b	Negative[a]	[M+Me-H]$^-$	[RCOO$_{sn1}$]$^-$ / [RCOO$_{sn2}$]$^-$	-200	-10	-60	-11	[f]
	PIP$_2$	2b	Negative[a]	[M+Me-H]$^-$	[RCOO$_{sn1}$]$^-$ / [RCOO$_{sn2}$]$^-$	-200	-10	-60	-11	[f]
	PS	2b	Negative	[M-H]$^-$	[RCOO$_{sn1}$]$^-$ / [RCOO$_{sn2}$]$^-$	-100	-10	-40	-10	(Tarazona *et al.* 2015)
Sphingolipids										

Class	Subclass		Ionization	Q1	Q3	CP	EP	CE	CXP	Reference
	LCB	2c	Positive	[M+H]$^+$	[M-H$_2$O+H]$^+$ / [M-2H$_2$O+H]$^+$ / [M-3H$_2$O+H]$^+$	50	10	20/ 25/ 30	10	(Tarazona *et al.* 2015)
Neutral lipids	Cer	2b	Positive	[M+H]$^+$	[LCB-2H$_2$O+H]$^+$	100	10	50	10	(Tarazona *et al.* 2015)
Glycolipids	HexCer	2b	Positive	[M+H]$^+$	[LCB-2H$_2$O+H]$^+$	120	10	55	10	(Tarazona *et al.* 2015)
	Hex-GlcA-IPC[g] HexN-GlcA-IPC HexNAc-GlcA-IPC	2b	Positive	[M+NH$_4$]$^+$	[Cer-2H$_2$O+H]$^+$	160	10	76	10	(Markham and Jaworski 2007)
	Hex-Hex-GlcA-IPC Hex-HexN-GlcA-IPC Hex-HexNAc-GlcA-IPC	2b	Negative	[M-2H]$^{2-}$	[373]$^-$ / [RCO-C$_3$H$_6$NO-H]$^-$	-160	-10	-44	-10	(Buré *et al.* 2011)
Phospholipids	Cer-P	2b	Negative[b]	[M+2Ac-H]$^-$	[M+Ac-H]$^-$ / [M-H$_2$O-H]$^-$	-100	-10	-50	-10	f
	LCB-P	2c	Negative[b]	[M+3Ac-H]$^-$ & [M+2Ac-H]$^-$	[M+2Ac-H]$^-$ / [M+Ac-H$_2$O-H]$^-$ & [M+Ac-H]$^-$ / [M-H$_2$O-H]$^-$	-100	-10	-31	-10	(Zienkiewicz *et al.* 2020)
Sterol lipids										
Glycolipids	SG	2b	Positive	[M+NH$_4$]$^+$	[Sterol-OH]$^+$	100	10	22	10	(Tarazona *et al.* 2015)
	ASG	2b	Positive	[M+NH$_4$]$^+$	[Sterol-OH]$^+$	100	10	28	10	(Tarazona *et al.* 2015)
Neutral lipids	SE	1	Positive	[M+NH$_4$]$^+$	[Sterol-OH]$^+$	140	10	22	6	(Tarazona *et al.* 2015)
	FS	2b	Positive[b]	[M+Ac+NH$_4$]$^+$	[Sterol-OH]$^+$	240	10	23	34	(Liebisch *et al.* 2006)

Details of the solvent gradient are described in Table 1. The ionization mode depicts the polarity of the nanoESI source. Q1 and Q3 indicate the parent and product ions, respectively. CP, EP, CE and CXP indicate the declustering potential, entrance potential, collision energy and cell exit potential for the molecular species of the corresponding lipid classes, respectively.

Abbreviations for lipid classes and subclasses: Ara, Arabidopside; ASG, acylated steryl glucoside; Cer, ceramide; Cer-P, ceramide phosphate; CL, cardiolipin; DAG, diacylglycerol; DGCC, diacyl-carboxyhydroxymethylcholine; DGDG, digalactosyldiacylglycerol; DGMG, digalactosylmonoacylglycerol; DGTA, diacylglyceryl-hydroxymethyltrimethyl-β-alanine; DGTS, diacylglyceryl-*O*-(*N,N,N*-trimethyl)-homoserine; FS, free sterol; GlcA, -glucuronic acid; Hex, hexosyl; HexCer, hexosylceramide; HexN, hexosaminyl; HexNAc, *N*-acetylhexosaminyl; IPC, inositol phosphoceramide; LCB, long-chain base; LCB-P, long-chain base phosphate; LPA, lysophosphatidic acid; LPC, lysophosphatidylcholine; LPE, lysophosphatidylethanolamine; LPG, lysophosphatidylglycerol; LPI, lysophosphatidylinositol; LPS, lysophosphatidylserine; PA, phosphatidic acid; PC, phosphatidylcholine; PE, phosphatidylethanolamine; PG, phosphatidylglycerol; PI, phosphatidylinositol; PIP, phosphatidylinositol phosphate; PIP$_2$, phosphatidylinositol bisphosphate; PS, phosphatidylserine; MGDG, monogalactosyldiacylglycerol; MGMG, monogalactosylmonoacylglycerol; SQDG, sulfoquinovosyldiacylglycerol; SE, steryl ester; SG, steryl glucoside; SQMG, sulfoquinovosylmonoacylglycerol; TAG, triacylglycerol

[a]Methylation or [b]acetylation before UPLC-nanoESI-MS/MS analysis

[c]Equals to [M-C$_6$H$_{11}$O$_6$+NH$_4$]$^+$

[d]Equals to [M-C$_{12}$H$_{21}$O$_{11}$+NH$_4$]$^+$

[e]Equals to [M-C$_6$H$_{11}$O$_5$-RCOO$_{Gal}$+NH$_4$]$^+$

[f]Unpublished data of the authors

[g]Nomenclature corresponding to (Fang *et al.* 2016)

3.8 Assembly of the Lipid Building Blocks for the Target Lipid List

To construct the mass transition lists, the identities of the following three lipid building blocks are either taken from the literature or determined experimentally by performing lipid measurements by e.g. GC or TLC using the biological sample.

Lipid building blocks:

1. Acyl chains (non-hydroxylated, hydroxylated)
2. Polar head groups (carboxyhydroxymethylcholine, dihexose, hexose, hexosyl glucosamine inositolphosphate (Hex-GlcA-Ins-P), hexosaminyl (HexN) GlcA-Ins-P, *N*-acetylhexosaminyl (HexNAc) GlcA-Ins-P, Hex-Hex-GlcA-Ins-P, Hex-HexN-GlcA-Ins-P, Hex-HexNAc-GlcA-Ins-P, hydroxymethyltrimethyl-β-alanine,(*N,N,N*-trimethyl)-homoserine, sulfoquinovose, phosphate, phosphocholine, phosphoethanolamine, phosphoglycerol, phosphoinositol, phosphoserine)
3. Backbones (glycerol, long-chain bases, sterols)

Classical lipid analytical methods are applied to analyze the building blocks as follows. Complex lipids are first partially disrupted by methanolysis under acidic or alkaline conditions (see Chapter xx the present book *Molecular Biology Plant Lipids*). Additional modifications of hydroxyl or amino groups via silylation (e.g., with BSTFA), and of methyl esters via transesterification into nitrogen-containing derivatives (e.g. pyrrolidides or 4,4-dimethyloxazolines) can be performed prior to the separation and detection of the analytes. Separation approaches including thin layer chromatography (TLC), high-performance liquid chromatography (HPLC) or gas chromatography (GC) can be coupled with detection methods ranging from staining procedures, flame ionization (FID) to mass spectrometry after electrospray ionization as well as electron impact ionization (for example: (Christie and Han 2010, Schneiter 2006), https://www.lipidhome.co.uk/, other chapters of the present book *Molecular Biology Plant Lipids*).

3.9 Data Analysis and Processing

1. Before starting data analysis, peak identification is supported by co-elution (same retention time and identical MS/MS patterns including head group-specific fragments) with analytical standards, authentic lipids isolated from biological extracts, and/or lipid data from the literature. Chromatographic rules for elution orders of the lipid classes and molecular species in a reversed-phase mode (structure of the head groups, length of the side chains, number of double bonds and hydroxyl groups) need to be considered (Figure 1) (Tarazona *et al.* 2015). The identities of molecular species can be confirmed by performing TLC

followed by lipid isolation from the plate and accurate mass measurements with a high-resolution mass spectrometer (time-of-flight analyzer, Orbitrap® analyzer etc.).

2. For peak integration with MultiQuant Software, set Gaussian smooth width to 2 points and minimal peak height to 300 points. Copy the entire output table from MultiQuant to Excel.

3. For data processing and lipid profiling, correct the raw peak area with the naturally occurring proportion of the ^{13}C isotopes. This contribution is calculated in correlation to the total number of carbon atoms from the respective molecular lipid species by multiplying the raw peak area by the correction factor α_{CN} (Iven *et al.* 2013). $\alpha_{CN} = 1 + 0.011\,n + 0.011^2\,n\,(n-1)/2$ (n = total number of carbon atoms of the lipid species). The ^{13}C isotope-corrected peak areas are then used to construct category-specific, class-specific, or subclass-specific lipid profiles.

3.10 Absolute Quantification of Lipid Subclasses by TLC Coupled with GC-FID

UPLC-nanoESI-MS/MS-based lipid analysis is a highly efficient method to separate molecular lipid species due to the robust reduction of sample matrix effects. One drawback of this chromatographic separation is the unequal solubility of the analytes along the elution gradient. Therefore, the ionization efficiency and fragmentation kinetics vary according to the target molecule. Different sample types, such as whole tissue extracts or membrane fractions, could limit the applicability of internal standards as well. Therefore, for absolute quantification of individual lipid subclasses, an analytical approach coupling TLC with GC/FID is used (Wang and Benning 2011).

1. The lipid extract is first separated by lipid class-specific TLC (for example: (Christie and Han 2010, Schneiter 2006), https://www.lipidhome.co.uk/, other chapters of the present book *Molecular Biology Plant Lipids*).

2. Lipid subclasses are stained reversibly with 0.05 % [*w/v*] primuline (Merck KGaA, Darmstadt, Germany) in 80 % [*v/v*] acetone(Kelly *et al.* 2013).

3. The lipids are isolated from the TLC plate.

4. Lipids are converted to fatty acid methyl esters (FAMEs) in methanol containing 5 % (*v/v*) sulfuric acid overnight at 110 °C (Cacas *et al.* 2016). For sterol quantification, the isolated lipids are silylated with *N,O*-bis(trimethylsilyl)trifluoroacetamide (BSTFA).

5. Pentadecanoic acid (15:0) is used as internal standard for glycerolipid quantification (Wang and Benning 2011), heptadecanoic acid (17:0) and 2-hydroxy-pentadecanoic acid for

sphingolipid quantification (Cacas *et al.* 2016),and cholestanol for sterol quantification (Wewer *et al.* 2011).

6. The resulting lipid subclass-specific FAMEs are analyzed by GC-FID and total amounts of the individual lipid subclasses are calculated based on normalization to internal standards (see chapter xx of the present book *Molecular Biology Plant Lipids*).

4 Notes

1. Wear glasses and cold-protective gloves when handling liquid nitrogen.
2. Cool down all equipment (mortar, pestle, mill jars, cups, tubes, spatula) with liquid nitrogen to avoid thawing and clogging of the sample on the equipment.
3. *Arabidopsis thaliana* rosettes are usually completely homogenized to fine powder using the Mixer Ball Mill MM200 for 1 min at 30 vibrations s^{-1}.
4. Wear gloves for sample extraction and sample handling before analysis.
5. Use pipette tips that are chemically resistant to organic solvents, and use glass tubes instead of plastic tubes for lipid extraction.
6. Lipids in organic solution and lipid extracts can be stored at -20 °C covered with argon for at least one year.

Chapter 3. Find the needle in the haystack – Analysis and identification of the rare and low-abundant ceramide phosphates in plants

The lipidomics platform described in chapter 2 includes the analysis of phosphorylated sphingolipids such as ceramide phosphates (Cer-Ps). The analysis of this trace lipid class based on a comprehensive method developmental process is explained within this chapter.

In plants, the overall sphingolipid biosynthesis takes place in the endomembranes (ER and Golgi apparatus); and the generation of Cer-P occurs presumably in both ER and Golgi apparatus where its synthesizing enzyme, ceramide kinase (CERK), locates (Bi *et al.* 2014) (Fig. 1). Cer-Ps are signaling molecules in several biological processes including the programmed cell death (PCD) and responses against abiotic and abiotic stresses (Bi *et al.* 2014, Brodersen *et al.* 2002, Donahue *et al.* 2010, Greenberg *et al.* 2000, Liang *et al.* 2003, Michaelson *et al.* 2016, Simanshu *et al.* 2014). The *Arabidopsis acd5* mutant displays an impaired activity of CERK, and exhibits an accelerated cell death (acd) phenotype (Dutilleul *et al.* 2015), hypersensitive responses when germinating at low temperature (Barrero-Sicilia *et al.* 2017, Dutilleul *et al.* 2015) and is more susceptible upon infection of *Pseudomonas syringae* with enhanced levels of reactive oxygen species (ROS) (Bi *et al.* 2014, Greenberg *et al.* 2000, Liang *et al.* 2003). In addition, *acd11*, which is mutated in the ACD11 (accelerated cell death 11) gene encoding a lipid transfer protein for Cer-Ps (Brodersen *et al.* 2002), displays spontaneous PCD responses driven by an altered sphingolipid homeostasis (Simanshu *et al.* 2014). To investigate the underlying regulatory mechanism and the biological functions of Cer-Ps, the analytical detection of the molecular Cer-P species is mandatory.

It has been demonstrated that Cer-P constitutes about 0.15 % of the total sphingolipids in human serum and plasma (Hammad *et al.* 2010); however, it accounts for very low abundance (presumably at the range of pmol/mg fresh weight) in plants (Simanshu *et al.* 2014). Based on the lipid profiles of other complex sphingolipids in plants, both Cer-Ps with non-hydroxylated and hydroxylated fatty acyl moieties (cCer-Ps and hCer-Ps) are considered to be present endogenously. Approaches with radiolabeled sphingolipids coupled with TLC separations and HPLC detections have been initiated to identify the molecular structures of endogenous Cer-Ps (Cantrel *et al.* 2011, Dutilleul *et al.* 2015). With this approach, chemical modifications including strong acidic hydrolysis to release the fatty acyl moieties and the derivatization of the amine group with o-phthaldialdehyde are required for detection by HPLCs. Although the radiolabeling approach enables the detection of trace analytes, the molecular species information of Cer-Ps, especially the fatty acyl moieties, is sacrificed. On the other hand, it is challenging to incorporate analysis of Cer-Ps in the LC-MS-based approaches due to the limited availability of authentic standards (only cCer-Ps can be obtained), the poor chromatographic separation and the trace abundance of endogenous Cer-Ps in the complex biological extracts. A couple of studies have pursued to analyze Cer-Ps via MS-based approaches (Bielawski *et al.* 2009, Markham and Jaworski 2007, Simanshu *et al.* 2014). They are based on a sphingolipid-specific extraction method and an intensive data processing based on calibration curves of non-phosphorylated ceramide standards in an artificial matrix. However, integration of the Cer-P analysis into the comprehensive lipidomics workflow would enable the interpretation of the Cer-P profiles in correlation to the entire lipidome of an organism.

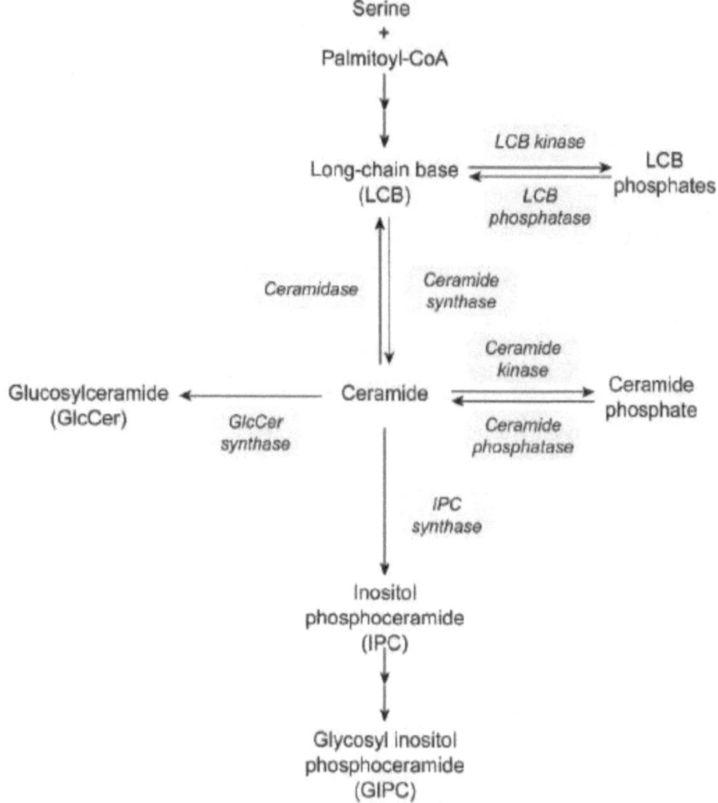

Figure 1. Sphingolipid biosynthesis. The *de novo* biosynthesis of long-chain bases (LCBs), long-chain base phosphates (LCB-Ps), ceramides (Cers), ceramide phosphates (Cer-Ps) and glucosylceramides (GlcCers) in the ER. Molecules of Cers are transported to the Golgi apparatus and undergo subsequent reactions in synthesizing complex sphingolipids including glycosyl inositol phosphoceramides (GIPCs) and distinct Cer-Ps.

In an attempt to accommodate Cer-Ps in the previously described LC-MS-based lipidomics workflow (Chapter 2), several steps in developing a compatible analytical method were made, including the establishment and optimization of the chromatographic separation, the MS detection by commercially available analytical Cer-P standards, and the extraction and purification of the endogenous Cer-Ps from *Arabidopsis* leaves.

It is challenging to separate the phosphorylated organic analytes via liquid chromatography. They adsorb to the capillaries of the electrospray under acidic condition, which is commonly applied to enhance the separation and resolution of biomolecules on C18 columns in reverse phase (RP)-LC systems. Moreover, stainless steel, which constitutes parts of the essential components on the path of the sample flow, can cause peak tailing of the phosphorylated analytes as well. Approaches such as optimizing the material of the capillaries (Fehrenbach and Wiese 2017), using monolithic columns (Tholey *et al.* 2005) or conducting the immobilized metal affinity chromatography (Krabbe *et al.* 2003, Mitulović and Mechtler 2006) have been applied on analyzing low-abundance phosphorylated analytes such as phosphopeptides. However, it is not feasible to employ these approaches in the presented established lipidomics workflow that contains a UPLC system with a C18 column and acidified (0.1 % acetic acid) mobile phases optimized for resolving the molecular species of most glycerolipids, sphingolipids and sterols.

Here, chemical modifications which can alter the phosphate groups of the analytes and thereby shield them from adsorbing to the analytical device were conducted to improve the chromatographic separation as well as the ionization efficiency of Cer-Ps. Modifications like methylation and acetylation have already been successfully incorporated to the lipidomics workflow when analyzing other phosphorylated lipids (phosphoinositides, phosphatidic acids and long-chain base phosphates (LCB-Ps)) (Chapter 2). The chemical modified analytical cCer-P standards were firstly monitored by direct infusion (DI)-MS to evaluate their ionization efficiencies (Fig. 2). Analyses of enhanced mass spectrum (EMS) scans in both positive and negative ion modes were conducted; however, ion signals were only detected in the negative ion mode. For the analytical standard cCer-P (18:1;2/16:0), the ion signal of [M-H]⁻ (*m/z* = 616.5 Da) was detected for the non-derivatized compound, [M+1Me-H]⁻ and [M+2Me-H]⁻ (*m/z* = 630.5 and 644.5 Da, respectively) for the methylated derivatives, and [M-H₂O-H]⁻, [M+1Ac-H]⁻ and [M+2Ac-H]⁻ (*m/z* = 598.4, 658.4 and 700.6 Da, respectively) for the acetylated derivatives. The mass spectra of the methylated compounds contained additionally an ion signal of an unknown structure (i.e. *m/z* = 649.2 Da). The ions derived from the acetylated derivatives, especially the bisacetylated ions [M+2Ac-H]⁻, provided the highest signals and greater signal-to-noise ratio in comparison to the others. The acetylated derivatives were therefore favored for conducting the following experiments.

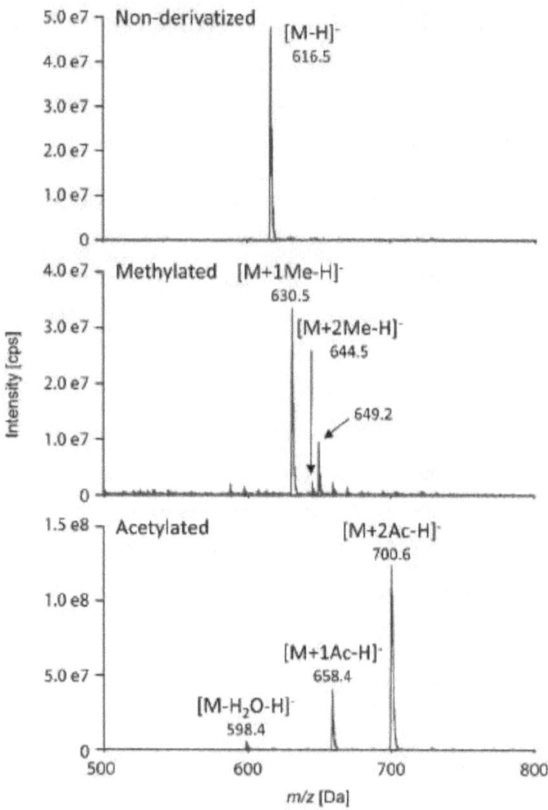

Figure 2. Mass spectrum of the analytical standard cCer-P (18:1;2/16:0) in the negative ion mode. Non-derivatized, methylated and acetylated analytical Cer-P standards were analyzed by enhanced mass spectrum (EMS) scans. Declustering potential = -100 V, collision energy = -10 V. The structure of the ion with m/z = 649.2 of the methylated derivatives is unknown.

To optimize the MS detection, the ion fragmentation of the bisacetylated Cer-P derivatives were investigated by DI-MS analyses in the negative ion mode (Fig. 3). The precursor ion [M+2Ac-H]⁻ of cCer-P (18:1;2/16:0) with m/z = 700.6 Da fragments into four product ions; (1) [M+1Ac-H]⁻ with m/z = 658.4 Da results from a loss of an acetyl group, (2) [M-H]⁻ with m/z = 616.3 results from the loss of two acetyl groups, (3) [M-H₂O-H]⁻ with m/z = 598.4 Da results from a loss of two acetyl groups and an additional water loss and (4) [LCB+PO₄]⁻ with m/z = 360.2 Da corresponds to the phosphorylated long-chain base (LCB) backbone. Since the ion fragment [LCB+PO₄]⁻ represents the exclusive structure-specific product ion, the collision energy (CE) was optimized to -60 V with respect to the maximal signal intensity of this ion mass.

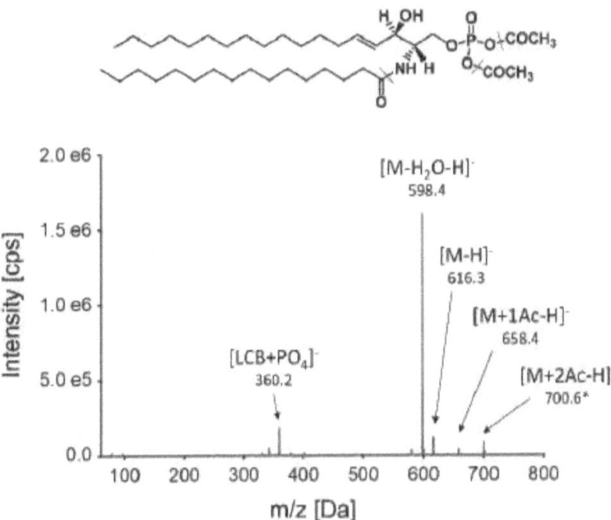

Figure 3. Fragmentation pattern and the product ion scan of the bisacetylated cCer-P (18:1;2/16:0) in the negative ion mode with collision energy = -60 V. The precursor ion (*) is [M+2Ac-H]⁻ (*m/z* = 700.6 Da).

In the next step, the Cer-P analysis was implemented into the LC-MS-based workflow using the multiple reaction monitoring (MRM) detection mode. Therefore, mass pairs of the respective precursor ion and the fragment ion - so called mass transitions - were generated to ensure and evaluate the chromatographic resolution of the analytical Cer-P standards (Fig.4). Equal concentrations (500 µM) of the analytical standard cCer-P (18:1;2/16:0) were measured directly and after both chemical derivatization procedures (methylation and acetylation) in the negative ion mode. As precursor ions, the deprotonated masses as well as the dehydrated masses of the non-derivatized, methylated (single, two times and three times) and acetylated ions (single, two times and three times), respectively, were analyzed. For all transitions, the backbone-specific fragment ion [LCB+PO₄]⁻ was used. Among all, only chromatographic signals of the mass transitions resulting from the deprotonated precursor ions of non-derivatized, monomethylated, monoacetylated and bisacetylated molecules were detectable. Although both derivatization methods enhanced the resolution of the LC-MS analysis, the acetylated standards provided the highest ion signals for the MS detection in comparison to the other compounds. Therefore, the acetylation procedure was selected for the further establishment of the LC-MS-based Cer-P detection method.

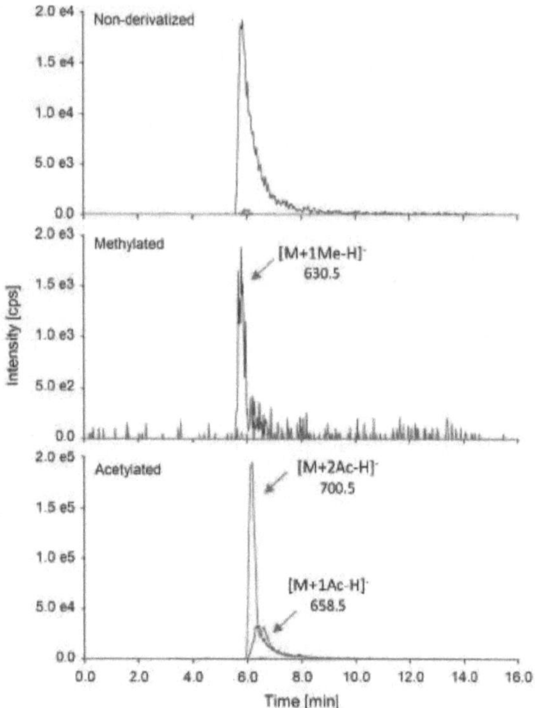

Figure 4. Derivatization of Cer-P standards improves the chromatographic resolution and the ionization efficiency. Non-derivatized, methylated and acetylated cCer-P (18:1;2/16:0) with equal concentration (500 µM) were measured in the negative ion mode. Ac: acetyl group; Me: methyl group. Fragments of [LCB+PO$_4$]⁻ (m/z = 360.2) derived from deprotonated and dehydrated precursors of the non-derivatized, methylated and acetylated molecules, respectively, were analyzed. Deprotonated and dehydrated precursors of non-derivatized (m/z = 616.5 and 598.5), monomethylated (m/z = 630.5 and 612.5), bismethylated (m/z = 644.5 and 626.5), trimethylated (m/z = 658.5 and 640.5), monoacetylated (m/z = 658.5 and 640.5), bisacetylated (m/z = 700.5 and 682.5) and triacetylated (m/z = 742.5 and 724.5) ions, respectively, were included into the MRM-based analysis.

Based on the fragmentation behavior of the analytical Cer-P standards, a MRM-based method was constructed to enable the analysis of the endogenous Cer-Ps in *Arabidopsis*. Therefore, all possible molecular Cer-P species were calculated based on the usage of 18:0;2, 18:1;2, 18:2;2, 18:0;3 or 18:1;3 as LCB backbones (C:DB;OH for numbers of carbon, double bond and hydroxyl group, respectively). These LCB backbones were paired with non- and monohydroxylated fatty acyl moieties of 16-28 carbons in length with zero or one double bond. This resulted in the prediction of 140 possible Cer-P species (Appendix. 1). For the analysis of their endogenous occurrence, all corresponding precursor ions representing the bisacetylated ions [M+2Act-H]⁻ were combined with three different fragment ions ([LCB+PO₄]⁻, [M+1Act-H]⁻ and [M-H₂O-H]⁻, respectively). The conducted MS parameters for all 420 transitions were as followed: -100 V declustering potential (DP), -10 V entrance potential (EP), -50 V CE and -10 V collision cell exit potential (CXP). For the chromatographic elution, the solvent gradient 2b was used as described in table 1 of Chapter 2. With this elution gradient, the bisacetylated cCer-P (18:1;2/16:0) and cCer-P (18:1;2/24:0) standards eluted at 6.3 and 9 min, respectively (Fig. 5). The intensity patterns of the compound-specific transitions correspond to the result of the DI-MS analysis. This was indicated by the highest signal intensities for the fragment ions [M-H₂O-H]⁻ followed by [M+1Ac-H]⁻ and the much lower intensity for [LCB+PO₄]⁻ even though the MS parameters were optimized of the detection of the backbone-specific fragments.

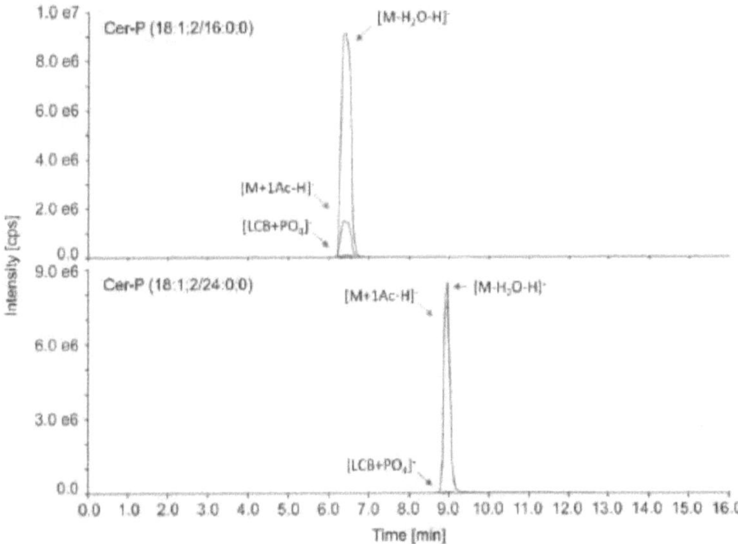

Figure 5. LC-MS analysis of the analytical standards cCer-P (18:1;2/16:0) and cCer-P (18:1;2/24:0) in the negative ion mode. The applied elution gradient starts from 65 % solvent B (tetrahydrofuran : methanol : 20 mM ammonium acetate (6:3:1, v.v.v) with 0.1 % acetic acid, v.v) and 35 % solvent A (methanol : 20 mM ammonium acetate (3:7, v.v) with 0.1 % acetic acid, v.v) corresponding to gradient 2b from Chapter 2. Chromatographic signals are labeled with the product ion structures of the respective MRM transitions. The precursor ions are the bisacetylated ions [M+2Act-H]⁻ of the respective molecules.

To evaluate the analytical quality of the established LC-MS method for the Cer-P analysis, the limits of detection and analysis were inspected by analyzing serial dilutions of the analytical Cer-P standards in the matrixes of pure organic solvent and *Arabidopsis* leaf extract, respectively (Fig. 6). Both limits of detection and analysis of cCer-P (18:1;2/16:0) and cCer-P (18:1;2/24:0) achieved femtomole (10^{-15} mol) as the lowest quantity, of which the signals can be distinguished confidently from the background noises. The limits of detection and analysis are lower than the presumed concentration of the endogenous Cer-Ps (< 1 pmol/mg fresh weight). The results correlate with the limits of detection and analysis for LCB-Ps (data not shown).

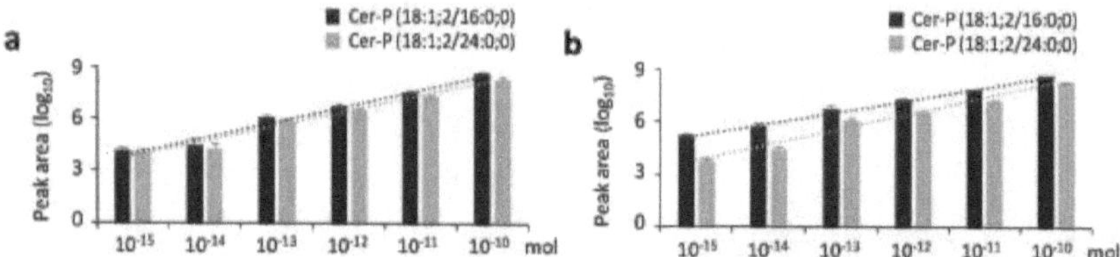

Figure 6. Limits of detection and analysis concerning Cer-Ps of the LC-MS method. Cer-P standards, cCer-P (18:1;2/16:0) and cCer-P (18:1;2/24:0), were analyzed after dilution in pure solvents to determine the limit of detection (a) and in the presence of leaf extracts (lipids resulting from 250 mg fresh weight ml^{-1}) to determine the limit of analysis (b), respectively. The analyses were conducted with three independently prepared standard solutions and each were analyzed three times under the respective conditions.

Since Cer-Ps represent only trace amounts in plants, the identification and structural verification of the endogenous Cer-Ps within complex lipid extracts is challenging. Therefore, in addition to the establishment of a sensitive LC-MS method, the endogenous Cer-Ps from *Arabidopsis* leaves had to be massively enriched. The extraction procedure of the established lipidomics workflow (Chapter 2) had already been optimized to efficiently extract a broad range of lipid classes, especially sphingolipids (Tarazona *et al.* 2015). Additionally, a TLC separation approach was established to enrich and purify the endogenous Cer-Ps from the total lipid extract. Several TLC methods targeting the separation of complex sphingolipids have been published hereto (Cacas *et al.* 2016, Dutilleul *et al.* 2015, Horibata *et al.* 2004, Lenarčič *et al.* 2017, Nakagawa *et al.* 1999, Park *et al.* 2014, Tanaka *et al.* 2013, Tidhar *et al.* 2015). The primary components of these TLC developing solutions comprise various amounts of chloroform and methanol. To adjust the solvent polarity for better separation of the analytes, calcium chloride or ammonia are commonly supplied. In this work, the addition of calcium chloride was avoided as its crystals may cause clogging or increasing the pressure in the following LC-MS procedure. Therefore, two TLC solvent mixtures were selected to evaluate their efficiency in separating the Cer-Ps from the other complex sphingolipids. Solvent mixture A – chloroform : methanol : 28% ammonia, 65:35:6 (v.v.v) (Tanaka *et al.* 2013) and solvent mixture B – chloroform : methanol : water, 60:35:8 (v.v.v) (Horibata *et al.* 2004) were conducted to separate a standard mixture composed of cCer (18:1;2/12:0), cCer-P (18:1;2/12:0), cGlcCer (18:1;2/12:0) and the most similar purchasable analogue of plant GIPCs, monosialotetrahexosyl ganglioside (GM1) (Fig. 7). The TLC separation was additionally inspected with respect to its efficiency to separate Cer-Ps from the glycerophospholipid phosphatidylcholine (PC) by using the analytical PC standard, PC (17:0/17:0). The main abundant lipid class PC can interfere with the sphingolipid detection during the later LC-MS analysis (Markham *et al.* 2006). Although both TLC solvent mixtures were capable of separating cCer-P (18:1;2/12:0) from the other selected standards. (Fig. 7, both lanes 1), the separation efficiency of solvent mixture B with respect to Cer-P and GM1 was more reliable.

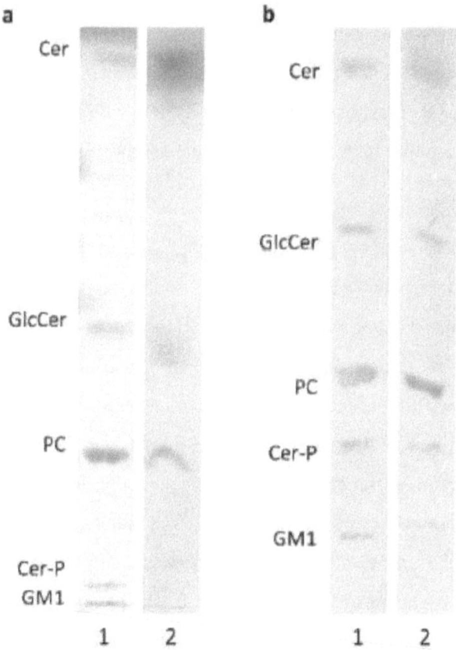

Figure 7. Separation of complex sphingolipid standards with chloroform : methanol : 28 % ammonia, 65:35:6 (v.v.v) (a) and chloroform : methanol : water, 60:35:8 (v.v.v) (b). (1) Standard mixture of 1 µg cCer (18:1;2/12:0), 1 µg cCer-P (18:1;2/12:0), 1 µg cGlcCer (18:1;2/12:0), 5 µg GM1 and 5 µg PC (17:0/17:0) were separated with the two different developing solvent systems. (2) The recovery of the complex sphingolipids was monitored by an additional TLC separation after re-extraction of the lipid spots with the respective TLC developing solutions. The lipid spots were visualized at 170 °C after incubating with copper sulfate solution.

To enable the structure analysis by LC-MS after the TLC separation, the Cer-P-corresponding lipid spot had to be re-extracted from the TLC silica gel. In order to investigate the recovery efficiency of the two TLC solvent mixtures, the lipid-containing silica was scraped off from the TLC plates, re-extracted with the same solvents as the developing solutions and redeveloped accordingly with either solvent mixture A or B on additional TLC plates (Fig. 7, lanes 2). The re-extraction procedure was optimized by incubating the silica-solvent mixtures at higher temperature (50 °C) for 30 min and with constant sonication for another 30 min to achieve higher recovery rates (data not shown). Although, the three analytical standards cCer (18:1;2/12:0), cGlcCer (18:1;2/12:0) and PC (17:0/17:0) were efficiently recovered with solvent mixture A, the analytical standards cCer-P (18:1;2/12:0) and GM1 were not detectable after the re-extraction by the subsequent TLC separation. In contrast to that, the usage of solvent mixture B resulted into high recovery of all five analytical standards, including Cer-P and GM1. In addition, the composition of solvent mixture B based on pure organic solvents (chloroform, methanol and water) without any salty or non-neutral consumables is more compatible with the following usage of a LC-MS device. Therefore, solvent mixture B was selected for Cer-P identification based on the sequential order of (1) lipid extraction, (2) TLC separation, (3) re-extraction of putative Cer-Ps from the silica gel of the TLC plate, (4) acetylation of the resulting lipids and (5) LC-MS analysis of molecular Cer-P species.

Like other sphingolipids, the endogenous Cer-Ps are presumed to be localized at the endomembranes (Michaelson *et al.* 2016). Therefore, instead of lipid extracts from the entire plant tissue, lipid extracts from the microsomes (MCs, crude total membrane fraction) as described in Chapter 2 were selected to identify the endogenous Cer-P species from *Arabidopsis* leaves. At first, lipid extracts from MCs were separated by TLC to investigate the resolution efficiency of the selected TLC system with respect to complex plant lipid extracts (Fig. 8, lane 2). For this, the lipid extract of the MCs was additionally spiked with the analytical standard mixture to inspect the resolution efficiency with respect to the low abundant endogenous Cer-Ps (Fig. 8, lane 1). The TLC separation resulted into a complete separation of all minor sphingolipid classes from the main abundant glycerophospholipids. However, the amount of endogenous Cer-Ps was under the detection limit of the copper sulfate staining (Fig. 8, lane 2). To improve the LC-MS-based analysis limit of the minor sphingolipids by avoiding analytical interference with the main abundant glycerophospholipids, the lipid extracts of the MCs were treated with methylamine that hydrolyzes the glycerolipids (Cacas *et al.* 2016, Markham *et al.* 2006). The resulting methylamine-treated MCs were additionally separated by TLC (Fig. 8, lane 3). The indicated area of the silica gel, which corresponds to the position of the putative endogenous Cer-Ps, was scraped off from the TLC plate after visualizing with primuline solution. The lipids were re-extracted with solvent mixture B from the silica gel, acetylated and finally analyzed by LC-MS.

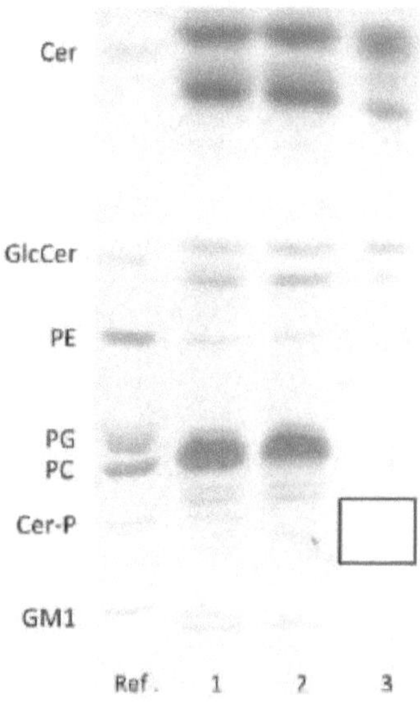

Figure 8. TLC separation of lipid extracts of the microsomal fraction (MC) of *Arabidopsis* leaves. (1) MC extract spiked with the standard mixture composed of 1 µg cCer (18:1;2/12:0), 1 µg cCer-P (18:1;2/12:0), 1 µg cGlcCer (18:1;2/12:0), 5 µg GM1 and 5 µg PC (17:0/17:0). (2) MC extract. (3) Methylamine-treated MC extract. The reference lane (Ref) includes 5 µg PE (17:0/17:0) and 5 µg PG (17:0/17:0) additionally to the standard mixture mentioned for (1). The developing solution B is composed of chloroform : methanol : water, 60:35:8 (v.v.v). The lipid spots were visualized at 170 °C after incubating with copper sulfate solution. For following LC-MS analysis, TLC plates were visualized with 0.5 mg ml^{-1} primuline in acetone : water (8:2, v.v) under 528 nm UV light and the targeted lipid spots (as labeled in lane 3) were scraped off from the TLC plate for subsequent re-extraction.

The TLC-purified and acetylated Cer-P fractions from the *Arabidopsis* MCs resulted in two chromatographic signals with the identical retention time of 6.2 min during the LC-MS analysis (Fig. 9). Both corresponding mass transitions result from the precursor ion with m/z = 816.4 Da paired with two different fragment ions ([M+1Act-H]⁻ with m/z = 774.4 Da and [M-H$_2$O-H]⁻ with m/z = 714.4 Da). Based on these masses of the precursor and fragment ions, both chromatographic signals may result from a monounsaturated Cer-P species composed of a trihydroxylated 18-cabon LCB backbone conjugated with a monohydroxylated 22-carbon fatty acyl moiety. Therefore, it was tentatively designated as hCer-P (18:0;3/22:1) or hCer (18:1;3/22:0), since the chromatographic signal of the corresponding mass transition based on the backbone-specific fragment ion was not detectable. Another chromatographic signal at 8.4 min probably results from the TLC silica gel since it can be detected in all samples including the blank extracts.

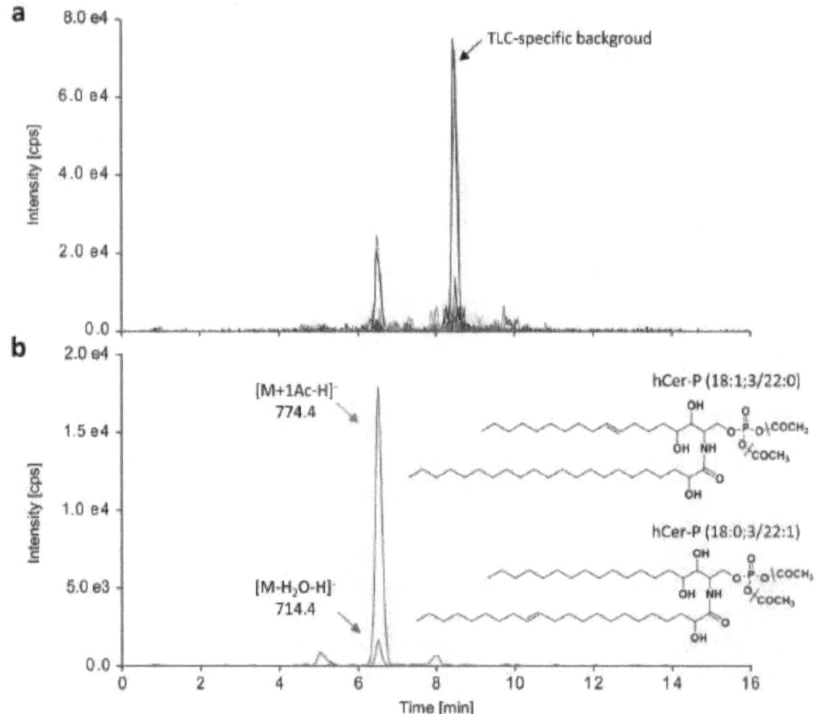

Figure 9. LC-MS analysis of the TLC-purified and acetylated Cer-P fraction from *Arabidopsis* MCs . The merged chromatogram of all analyzed MRM transitions resulting from the TLC re-extracted sample (a) and the chromatogram of the extracted transitions including the fragments with the loss of one acetyl group [M+1Act-H]⁻ (m/z = 774.4 Da) and the loss of two acetyl groups plus one water molecule [M-H₂O-H]⁻ (m/z = 714.4 Da) from the bisacetylated precursor (m/z = 816.4 Da), which corresponds to the indicated structure of the putative Cer-P (b). The solvent gradient corresponds to the conditions as described in Fig. 5 and the parameter for the MS detection are shown in Appendix 1.

Since the backbone-specific fragment ion of the endogenous Cer-P species was not detectable, the structure of this compound cannot be entirely resolved. According to the *de novo* sphingolipid biosynthesis pathway (Fig. 1), Cer-Ps are converted from Cers by CERK. Hence, the Cer profile of the *Arabidopsis* MCs was utilized as a reference to evaluate the structure of the putative Cer-P species.

Based on the Cer species profile, the putative species were predicted as hCer-P (18:1;3/22:0) due to the significantly higher levels of the hCer (18:1;3/22:0) in comparison to hCer (18:0;3/22:1) in the MCs (Fig. 10).

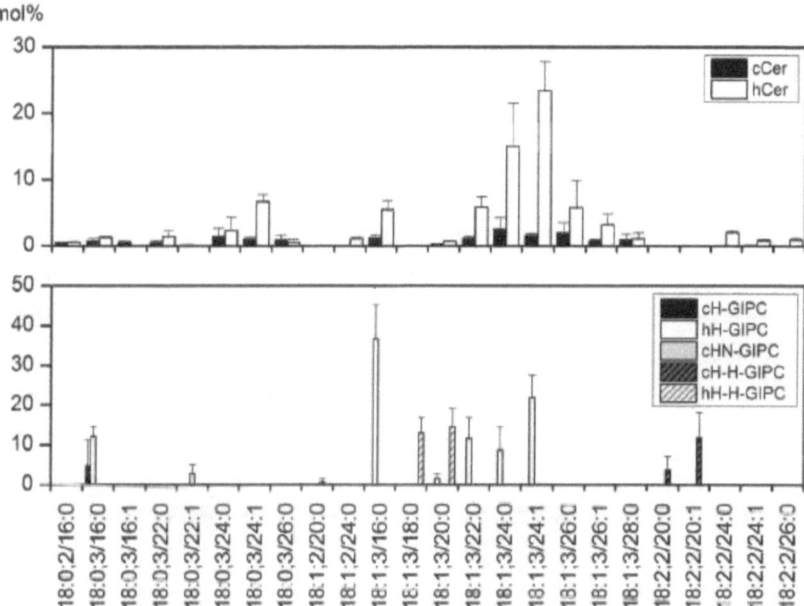

Figure 10. Ceramides, series A and B GIPCs from *Arabidopsis* MCs were analyzed by LC-MS. Data represent the mean value in mol % of an individual lipid species in either overall ceramides or GIPCs from three independent experiments ± SD. Non-hydroxylated and hydroxylated varieties (c- and h-) of ceramides (Cer) as well as series A GIPCs (H-GIPC and HN-GIPC) and series B GIPCs (H-H-GIPC) are illustrated. GIPC, glycosyl inositol phosphoceramides; H, hexosyl; HN, N-acetylhexosaminyl.

In plants, Cer-Ps can be synthesized alternatively via hydrolyzing GIPCs by GIPC-specific phospholipase D (GIPC-PLD). In cabbage (*Brassica oleracea*), the resulting Cer-Ps contain a 18:1;3 LCB backbone with various α-hydroxylated fatty acyl moieties (Hasi *et al.* 2019a, Hasi *et al.* 2019b, Tanaka *et al.* 2013). Mild activity of GIPC-PLD has also been identified in fractions of *Arabidopsis* crude protein extract; however, the endogenous products of the *Arabidopsis* GIPC-PLD were not profiled hereto. Nevertheless, according to the GIPC profile of the *Arabidopsis* MCs, the relative amount of hH-GIPC (18:1;3/22:0) is higher than that of hH-GIPC (18:0;3/22:1) (Fig. 10). Therefore, hCer-P (18:1;3/22:0) is suggested to be the molecular structure according to both profiles of Cers and GIPCs.

In this study, several developments have been conducted to identify the endogenous Cer-Ps in *Arabidopsis*. In order to obtain their backbone-specific information and to incorporate the analysis of Cer-Ps in the LC-MS-based lipidomics platform, (1) the fragmentation parameters in the MS system were specifically optimized to identify the backbone-specific signals of Cer-Ps, (2) the matrix effects from the abundant glycerophospholipids in leaf extracts were eliminated via methylamine treatment and TLC purification, and (3) the concentration of the analytes was enhanced as well by minimizing the sample volume before LC-MS analyses. Nevertheless, further investigations of the endogenous Cer-Ps are required to verify their structure and understand their functions in plants. For instance, beside the presented approach, which links the TLC purification of plant Cer-Ps with LC-MS profiling, an additional separation by collecting the MS signal-specific eluents from the LC system can be conducted. The resulting fractions that contain the endogenous Cer-Ps can be subjected to DI-MS analysis, which allows dynamic tuning of the MS parameters for determining the optimized parameters of individual Cer-P species. Lipid extractions of stress-challenged *Arabidopsis* (i.e. cold stress), lipid biosynthesis mutants (i.e. CERK overexpression line) or from other plant species (i.e. cabbage) and organisms (i.e. human plasma) may obtain higher amount of endogenous Cer-Ps for developing further analytical methods. Implementing the presented approaches creates a great potential in establishing a global lipidomics analyzing system which includes species information of endogenous Cer-Ps in the LC-MS-based lipidomics workflow.

Materials and methods

Thin layer chromatographic separation of complex sphingolipids

Lipids extracted from microsomes (MCs) were spotted onto TLC 60 plates (20 x 20 cm^2, Merck KGaA, Darmstadt, Germany) in parallel with complex sphingolipid standards (Sigma-Aldrich, Taufkirchen, Germany). Solvent mixtures of chloroform : methanol : 28 % ammonia (65:35:6, v.v.v) (Tanaka *et al.* 2013) and chloroform : methanol : water (60:35:8, v.v.v) (Horibata *et al.* 2004) were used as developing solution for method A and B, respectively. To continue with lipid re-extraction, lipid spots on TLC plates were visualized after spraying with 0.5 mg ml^{-1} primuline in the mixture of acetone : water (8:2, v.v) (White *et al.* 1998) under UV light at 528 nm wavelength. Alternatively, the lipids on the TLC plate can be stained irreversibly by immersing in copper sulphate solution (0.4 M $CuSO_4$ in 6.8 % (v.v) H_3PO_4), followed by drying at 100 °C and visualizing at 170 °C.

Lipid re-extraction from the TLC plate

The lipid-containing silica powder scraped off from the corresponding TLC area were incubated with the same solvent mixture as the respective developing solution at 50 °C for 30 min followed by continuous sonication for another 30 min. After centrifugation at 800 g for 20 min, the clear supernatant was transferred to clean tubes and evaporated under streams of nitrogen gas until dryness. Lipids were reconstituted by different solvents according to the following experimental procedure.

Direct infusion-mass spectrometry

Acetylated standards were reconstituted to 1 µM in tetrahydrofuran : methanol : water (4:4:1, v.v). An aliquot of 5 µl sample was automatically loaded and delivered to the chip-based nano-electrospray of TriVersa Nanomate (Advion BioScience, Ithaca, NY, USA). The MS analysis was carried out on a 6500 QTRAP tandem mass spectrometer (AB Sciex, Framingham, MA, USA). The ionization source parameters were set as curtain gas at 20, ion spray voltage at -1.45 kV and declustering potential at -100 V. The data acquisition was monitored by Analyst software (AB Sciex). The optimized collision energy for the bisacetylated Cer-P standards with fatty acyl chain with 16 to 24 carbons in length were in the range between -45 and -60 V.

Appendix 1. Targeted LC-MS-based analysis of putative *Arabidopsis* Cer-Ps. MRM list includes all possible endogenous species comprised of 18:0;2, 18:1;2, 18:2;2, 18:0;3 and 18:1;3 LCB backbones and fatty acyl moieties with 16-24 carbons in length, maximum one double bond and one hydroxyl group. All precursors are bisacetylated [M+2Act-H]⁻; the fragment ions of the phosphorylated backbone [LCB+PO₄]⁻, from the loss of one acetyl group [M+1Act-H]⁻ and from the loss of both acetyl group plus one water molecule [M-H₂O-H]⁻ were analyzed in the acquisition method.

Precursor ion (*m/z*)		Fragment ions (*m/z*)			MS parameter (V)			
Name	[M+2Act-H]⁻	[LCB+PO₄]⁻	[M+1Act-H]⁻	[M-H₂O-H]⁻	DP	EP	CE	CXP
(18:0;2/16:0;0)P+2Ac	702,3	362,2	660,3	600,3	-100	-10	-50	-10
(18:0;2/16:0;1)P+2Ac	718,3	362,2	676,3	616,3	-100	-10	-50	-10
(18:0;2/16:1;0)P+2Ac	700,3	362,2	658,3	598,3	-100	-10	-50	-10
(18:0;2/16:1;1)P+2Ac	716,3	362,2	674,3	614,3	-100	-10	-50	-10
(18:0;2/18:0;0)P+2Ac	730,4	362,2	688,4	628,4	-100	-10	-50	-10
(18:0;2/18:0;1)P+2Ac	746,4	362,2	704,4	644,4	-100	-10	-50	-10
(18:0;2/18:1;0)P+2Ac	728,4	362,2	686,4	626,4	-100	-10	-50	-10
(18:0;2/18:1;1)P+2Ac	744,3	362,2	702,3	642,3	-100	-10	-50	-10
(18:0;2/20:0;0)P+2Ac	758,4	362,2	716,4	656,4	-100	-10	-50	-10
(18:0;2/20:0;1)P+2Ac	774,4	362,2	732,4	672,4	-100	-10	-50	-10
(18:0;2/20:1;0)P+2Ac	756,4	362,2	714,4	654,4	-100	-10	-50	-10
(18:0;2/20:1;1)P+2Ac	772,4	362,2	730,4	670,4	-100	-10	-50	-10
(18:0;2/22:0;0)P+2Ac	786,4	362,2	744,4	684,4	-100	-10	-50	-10
(18:0;2/22:0;1)P+2Ac	802,4	362,2	760,4	700,4	-100	-10	-50	-10
(18:0;2/22:1;0)P+2Ac	784,4	362,2	742,4	682,4	-100	-10	-50	-10
(18:0;2/22:1;1)P+2Ac	800,4	362,2	758,4	698,4	-100	-10	-50	-10
(18:0;2/24:0;0)P+2Ac	814,4	362,2	772,4	712,4	-100	-10	-50	-10
(18:0;2/24:0;1)P+2Ac	830,4	362,2	788,4	728,4	-100	-10	-50	-10
(18:0;2/24:1;0)P+2Ac	812,4	362,2	770,4	710,4	-100	-10	-50	-10
(18:0;2/24:1;1)P+2Ac	828,4	362,2	786,4	726,4	-100	-10	-50	-10
(18:0;2/26:0;0)P+2Ac	842,5	362,2	800,5	740,5	-100	-10	-50	-10
(18:0;2/26:0;1)P+2Ac	858,5	362,2	816,5	756,5	-100	-10	-50	-10
(18:0;2/26:1;0)P+2Ac	840,5	362,2	798,5	738,5	-100	-10	-50	-10
(18:0;2/26:1;1)P+2Ac	856,4	362,2	814,4	754,4	-100	-10	-50	-10
(18:0;2/28:0;0)P+2Ac	870,5	362,2	828,5	768,5	-100	-10	-50	-10
(18:0;2/28:0;1)P+2Ac	886,5	362,2	844,5	784,5	-100	-10	-50	-10
(18:0;2/28:1;0)P+2Ac	868,5	362,2	826,5	766,5	-100	-10	-50	-10
(18:0;2/28:1;1)P+2Ac	884,5	362,2	842,5	782,5	-100	-10	-50	-10
(18:1;2/16:0;0)P+2Ac	700,3	360,2	658,3	598,3	-100	-10	-50	-10
(18:1;2/16:0;1)P+2Ac	716,3	360,2	674,3	614,3	-100	-10	-50	-10
(18:1;2/16:1;0)P+2Ac	698,3	360,2	656,3	596,3	-100	-10	-50	-10
(18:1;2/16:1;1)P+2Ac	714,5	360,2	672,5	612,4	-100	-10	-50	-10
(18:1;2/18:0;0)P+2Ac	728,4	360,2	686,4	626,4	-100	-10	-50	-10
(18:1;2/18:0;1)P+2Ac	744,3	360,2	702,3	642,3	-100	-10	-50	-10
(18:1;2/18:1;0)P+2Ac	726,3	360,2	684,3	624,3	-100	-10	-50	-10
(18:1;2/18:1;1)P+2Ac	742,3	360,2	700,3	640,3	-100	-10	-50	-10
(18:1;2/20:0;0)P+2Ac	756,4	360,2	714,4	654,4	-100	-10	-50	-10
(18:1;2/20:0;1)P+2Ac	772,4	360,2	730,4	670,4	-100	-10	-50	-10
(18:1;2/20:1;0)P+2Ac	754,4	360,2	712,4	652,4	-100	-10	-50	-10
(18:1;2/20:1;1)P+2Ac	770,4	360,2	728,4	668,4	-100	-10	-50	-10
(18:1;2/22:0;0)P+2Ac	784,4	360,2	742,4	682,4	-100	-10	-50	-10
(18:1;2/22:0;1)P+2Ac	800,4	360,2	758,4	698,4	-100	-10	-50	-10
(18:1;2/22:1;0)P+2Ac	782,4	360,2	740,4	680,4	-100	-10	-50	-10

(18:1;2/22:1;1)P+2Ac	798,4	360,2	756,4	696,4	-100	-10	-50	-10
(18:1;2/24:0;0)P+2Ac	812,4	360,2	770,4	710,4	-100	-10	-50	-10
(18:1;2/24:0;1)P+2Ac	828,4	360,2	786,4	726,4	-100	-10	-50	-10
(18:1;2/24:1;0)P+2Ac	810,4	360,2	768,4	708,4	-100	-10	-50	-10
(18:1;2/24:1;1)P+2Ac	826,6	360,2	784,6	724,5	-100	-10	-50	-10
(18:1;2/26:0;0)P+2Ac	840,5	360,2	798,5	738,5	-100	-10	-50	-10
(18:1;2/26:0;1)P+2Ac	856,4	360,2	814,4	754,4	-100	-10	-50	-10
(18:1;2/26:1;0)P+2Ac	838,4	360,2	796,4	736,4	-100	-10	-50	-10
(18:1;2/26:1;1)P+2Ac	854,4	360,2	812,4	752,4	-100	-10	-50	-10
(18:1;2/28:0;0)P+2Ac	868,5	360,2	826,5	766,5	-100	-10	-50	-10
(18:1;2/28:0;1)P+2Ac	884,5	360,2	842,5	782,5	-100	-10	-50	-10
(18:1;2/28:1;0)P+2Ac	866,5	360,2	824,5	764,5	-100	-10	-50	-10
(18:1;2/28:1;1)P+2Ac	882,5	360,2	840,5	780,5	-100	-10	-50	-10
(18:2;2/16:0;0)P+2Ac	698,3	358,2	656,3	596,3	-100	-10	-50	-10
(18:2;2/16:0;1)P+2Ac	714,5	358,2	672,5	612,4	-100	-10	-50	-10
(18:2;2/16:1;0)P+2Ac	696,3	358,2	654,3	594,3	-100	-10	-50	-10
(18:2;2/16:1;1)P+2Ac	712,5	358,2	670,5	610,4	-100	-10	-50	-10
(18:2;2/18:0;0)P+2Ac	726,3	358,2	684,3	624,3	-100	-10	-50	-10
(18:2;2/18:0;1)P+2Ac	742,3	358,2	700,3	640,3	-100	-10	-50	-10
(18:2;2/18:1;0)P+2Ac	724,3	358,2	682,3	622,3	-100	-10	-50	-10
(18:2;2/18:1;1)P+2Ac	740,5	358,2	698,5	638,4	-100	-10	-50	-10
(18:2;2/20:0;0)P+2Ac	754,4	358,2	712,4	652,4	-100	-10	-50	-10
(18:2;2/20:0;1)P+2Ac	770,4	358,2	728,4	668,4	-100	-10	-50	-10
(18:2;2/20:1;0)P+2Ac	752,4	358,2	710,4	650,4	-100	-10	-50	-10
(18:2;2/20:1;1)P+2Ac	768,3	358,2	726,3	666,3	-100	-10	-50	-10
(18:2;2/22:0;0)P+2Ac	782,4	358,2	740,4	680,4	-100	-10	-50	-10
(18:2;2/22:0;1)P+2Ac	798,4	358,2	756,4	696,4	-100	-10	-50	-10
(18:2;2/22:1;0)P+2Ac	780,4	358,2	738,4	678,4	-100	-10	-50	-10
(18:2;2/22:1;1)P+2Ac	796,4	358,2	754,4	694,4	-100	-10	-50	-10
(18:2;2/24:0;0)P+2Ac	810,4	358,2	768,4	708,4	-100	-10	-50	-10
(18:2;2/24:0;1)P+2Ac	826,6	358,2	784,6	724,5	-100	-10	-50	-10
(18:2;2/24:1;0)P+2Ac	808,4	358,2	766,4	706,4	-100	-10	-50	-10
(18:2;2/24:1;1)P+2Ac	824,4	358,2	782,4	722,4	-100	-10	-50	-10
(18:2;2/26:0;0)P+2Ac	838,4	358,2	796,4	736,4	-100	-10	-50	-10
(18:2;2/26:0;1)P+2Ac	854,4	358,2	812,4	752,4	-100	-10	-50	-10
(18:2;2/26:1;0)P+2Ac	836,4	358,2	794,4	734,4	-100	-10	-50	-10
(18:2;2/26:1;1)P+2Ac	852,4	358,2	810,4	750,4	-100	-10	-50	-10
(18:2;2/28:0;0)P+2Ac	866,5	358,2	824,5	764,5	-100	-10	-50	-10
(18:2;2/28:0;1)P+2Ac	882,5	358,2	840,5	780,5	-100	-10	-50	-10
(18:2;2/28:1;0)P+2Ac	864,5	358,2	822,5	762,5	-100	-10	-50	-10
(18:2;2/28:1;1)P+2Ac	880,4	358,2	838,4	778,4	-100	-10	-50	-10
(18:0;3/16:0;0)P+2Ac	718,3	378,2	676,3	616,3	-100	-10	-50	-10
(18:0;3/16:0;1)P+2Ac	734,3	378,2	692,3	632,3	-100	-10	-50	-10
(18:0;3/16:1;0)P+2Ac	716,3	378,2	674,3	614,3	-100	-10	-50	-10
(18:0;3/16:1;1)P+2Ac	732,3	378,2	690,3	630,3	-100	-10	-50	-10
(18:0;3/18:0;0)P+2Ac	746,4	378,2	704,4	644,4	-100	-10	-50	-10
(18:0;3/18:0;1)P+2Ac	762,4	378,2	720,4	660,4	-100	-10	-50	-10
(18:0;3/18:1;0)P+2Ac	744,3	378,2	702,3	642,3	-100	-10	-50	-10
(18:0;3/18:1;1)P+2Ac	760,3	378,2	718,3	658,3	-100	-10	-50	-10
(18:0;3/20:0;0)P+2Ac	774,4	378,2	732,4	672,4	-100	-10	-50	-10
(18:0;3/20:0;1)P+2Ac	790,4	378,2	748,4	688,4	-100	-10	-50	-10
(18:0;3/20:1;0)P+2Ac	772,4	378,2	730,4	670,4	-100	-10	-50	-10

(18:0;3/20:1;1)P+2Ac	788,4	378,2	746,4	686,4	-100	-10	-50	-10
(18:0;3/22:0;0)P+2Ac	802,4	378,2	760,4	700,4	-100	-10	-50	-10
(18:0;3/22:0;1)P+2Ac	818,4	378,2	776,4	716,4	-100	-10	-50	-10
(18:0;3/22:1;0)P+2Ac	800,4	378,2	758,4	698,4	-100	-10	-50	-10
(18:0;3/22:1;1)P+2Ac	816,4	378,2	774,4	714,4	-100	-10	-50	-10
(18:0;3/24:0;0)P+2Ac	830,4	378,2	788,4	728,4	-100	-10	-50	-10
(18:0;3/24:0;1)P+2Ac	846,4	378,2	804,4	744,4	-100	-10	-50	-10
(18:0;3/24:1;0)P+2Ac	828,4	378,2	786,4	726,4	-100	-10	-50	-10
(18:0;3/24:1;1)P+2Ac	844,4	378,2	802,4	742,4	-100	-10	-50	-10
(18:0;3/26:0;0)P+2Ac	858,5	378,2	816,5	756,5	-100	-10	-50	-10
(18:0;3/26:0;1)P+2Ac	874,5	378,2	832,5	772,5	-100	-10	-50	-10
(18:0;3/26:1;0)P+2Ac	856,4	378,2	814,4	754,4	-100	-10	-50	-10
(18:0;3/26:1;1)P+2Ac	872,4	378,2	830,4	770,4	-100	-10	-50	-10
(18:0;3/28:0;0)P+2Ac	886,5	378,2	844,5	784,5	-100	-10	-50	-10
(18:0;3/28:0;1)P+2Ac	902,5	378,2	860,5	800,5	-100	-10	-50	-10
(18:0;3/28:1;0)P+2Ac	884,5	378,2	842,5	782,5	-100	-10	-50	-10
(18:0;3/28:1;1)P+2Ac	900,5	378,2	858,5	798,5	-100	-10	-50	-10
(18:1;3/16:0;0)P+2Ac	716,3	376,2	674,3	614,3	-100	-10	-50	-10
(18:1;3/16:0;1)P+2Ac	732,3	376,2	690,3	630,3	-100	-10	-50	-10
(18:1;3/16:1;0)P+2Ac	714,3	376,2	672,3	612,3	-100	-10	-50	-10
(18:1;3/16:1;1)P+2Ac	730,5	376,2	688,5	628,5	-100	-10	-50	-10
(18:1;3/18:0;0)P+2Ac	744,3	376,2	702,3	642,3	-100	-10	-50	-10
(18:1;3/18:0;1)P+2Ac	760,3	376,2	718,3	658,3	-100	-10	-50	-10
(18:1;3/18:1;0)P+2Ac	742,3	376,2	700,3	640,3	-100	-10	-50	-10
(18:1;3/18:1;1)P+2Ac	758,3	376,2	716,3	656,3	-100	-10	-50	-10
(18:1;3/20:0;0)P+2Ac	772,4	376,2	730,4	670,4	-100	-10	-50	-10
(18:1;3/20:0;1)P+2Ac	788,4	376,2	746,4	686,4	-100	-10	-50	-10
(18:1;3/20:1;0)P+2Ac	770,4	376,2	728,4	668,4	-100	-10	-50	-10
(18:1;3/20:1;1)P+2Ac	786,4	376,2	744,4	684,4	-100	-10	-50	-10
(18:1;3/22:0;0)P+2Ac	800,4	376,2	758,4	698,4	-100	-10	-50	-10
(18:1;3/22:0;1)P+2Ac	816,4	376,2	774,4	714,4	-100	-10	-50	-10
(18:1;3/22:1;0)P+2Ac	798,4	376,2	756,4	696,4	-100	-10	-50	-10
(18:1;3/22:1;1)P+2Ac	814,4	376,2	772,4	712,4	-100	-10	-50	-10
(18:1;3/24:0;0)P+2Ac	828,4	376,2	786,4	726,4	-100	-10	-50	-10
(18:1;3/24:0;1)P+2Ac	844,4	376,2	802,4	742,4	-100	-10	-50	-10
(18:1;3/24:1;0)P+2Ac	826,4	376,2	784,4	724,4	-100	-10	-50	-10
(18:1;3/24:1;1)P+2Ac	842,6	376,2	800,6	740,6	-100	-10	-50	-10
(18:1;3/26:0;0)P+2Ac	856,4	376,2	814,4	754,4	-100	-10	-50	-10
(18:1;3/26:0;1)P+2Ac	872,4	376,2	830,4	770,4	-100	-10	-50	-10
(18:1;3/26:1;0)P+2Ac	854,4	376,2	812,4	752,4	-100	-10	-50	-10
(18:1;3/26:1;1)P+2Ac	870,4	376,2	828,4	768,4	-100	-10	-50	-10
(18:1;3/28:0;0)P+2Ac	884,5	376,2	842,5	782,5	-100	-10	-50	-10
(18:1;3/28:0;1)P+2Ac	900,5	376,2	858,5	798,5	-100	-10	-50	-10
(18:1;3/28:1;0)P+2Ac	882,5	376,2	840,5	780,5	-100	-10	-50	-10
(18:1;3/28:1;1)P+2Ac	898,5	376,2	856,5	796,5	-100	-10	-50	-10

Chapter 4.

Loss of sphingolipid fatty acid α-hydroxylases triggers similar effects on the lipid composition of the plasma membrane as cold acclimation

The article is prepared for submission. The supplemental materials are attached at the end of the chapter.

Author contribution

Yi-Tse Liu designed all experiments. She performed the isolation of plasma membranes, extraction of membrane lipids, analysis and subsequent data processing of the TLC-GC and LC-MS/MS results. Moreover, she conducted the protein sample preparation and the subsequent data processing after the proteomics analysis. She analyzed, processed, displayed and discussed the data resulting from those experiments, and wrote the first version of the manuscript.

Loss of sphingolipid fatty acid α-hydroxylases triggers similar effects on the lipid composition of the plasma membrane as cold acclimation

Yi-Tse Liu[1], Cornelia Herrfurth[1,2], Kerstin Schmitt[3], Oliver Valerius[3], Ivo Feussner[1,2,4,*]

[1]University of Goettingen, Albrecht-von-Haller-Institute for Plant Sciences, Department of Plant Biochemistry, 37077 Goettingen, Germany

[2]University of Goettingen, Goettingen Center for Molecular Biosciences (GZMB), Service Unit for Metabolomics and Lipidomics, 37077 Goettingen, Germany

[3]University of Goettingen, Institute for Microbiology and Genetics, Department of Molecular Microbiology and Genetics, 37077 Goettingen, Germany

[4]University of Goettingen, Goettingen Center for Molecular Biosciences (GZMB), Department of Plant Biochemistry, 37077 Goettingen, Germany

*Correspondence: Ivo Feussner, Justus-von-Liebig-Weg 11, 37077 Goettingen, Germany; Fax:+49-551-39-25749; Email: ifeussn@uni-goettingen.de

Running Title: Effects of sphingolipid biosynthetic defects on the lipid composition of *Arabidopsis* plasma membrane

Keywords: *Arabidopsis thaliana*, cold, fatty acid α-hydroxylation, lipidomics, plasma membrane, sphingolipids

Significance statement

Maintaining the integrity of the plasma membrane (PM) under various environmental stresses is essential for the survival of plants. Detailed lipid profiling of PMs isolated from *Arabidopsis* wild type and a sphingolipid fatty acid α-hydroxylase mutant was performed to investigate the function of α-hydroxylated sphingolipids on the PM organization. Together with the analysis of transversal lipid distribution, the lipid landscape of the plant PM is outlined concerning the composition of lipid class and lipid species on the two membrane leaflets.

Introduction

Plants, being immobile organisms, are constantly challenged by various biotic and abiotic stresses. Therefore, the plasma membrane (PM), which represents the barrier between the cell and the outer environment, plays a critical role for their survival. PMs ought to adapt dynamically to the environmental changes such as the alterations of temperature and salinity, or the encounter of pathogens depending on the cell type and the developmental stage (Mamode Cassim *et al.* 2019, Niu and Xiang 2018). Many researches have focused on characterizing PM-localized or PM-associated proteins that take part in signal recognition and transduction, linking the external stimuli and intracellular signaling pathways (Jaillais and Ott 2020, Luschnig and Vert 2014). Large numbers of proteins involved in cellular metabolism, cell structure and traffic, protein maturation and turnover have been identified in the *Arabidopsis* PM proteome (Alexandersson *et al.* 2004, Marmagne *et al.* 2007, Marmagne *et al.* 2004). Although PM lipids play essential roles in establishing cell polarity, determining membrane fluidity as well as regulating enzyme activity and signal transduction (Grosjean *et al.* 2018, Guo *et al.* 2019, Hou *et al.* 2016, Zauber *et al.* 2014), our understanding concerning the lipid composition of the plant PM and its transversal distribution is rather limited.

Arabidopsis wild-type (WT) leaves contain mostly phospholipids and sterols in their PMs (47 % and 46 %, respectively) (Uemura *et al.* 1995). Although sphingolipids represent only a small portion (7 %), they are essential functional components and play decisive roles in modulating the membrane biophysical properties and mediating the plant defense responses (Berkey *et al.* 2012, Huby *et al.* 2020, Michaelson *et al.* 2016). The core structure of sphingolipids is a long-chain base (LCB) which can be further converted to ceramide (Cer) by *N*-acylation. Subsequent addition of glucose or hexosyl inositol phosphates to Cer generates so-called complex sphingolipids, namely the glucosylceramides (GlcCers) and the glycosyl inositol phosphoceramides (GIPCs). Overall analysis of the sphingolipid constituents in *Arabidopsis* PM has demonstrated that the majority contains trihydroxylated LCBs and fatty acyl chains with 24 carbons in length (Liu *et al.* 2020). In addition, the complex sphingolipids – GIPCs, which can function as pathogenic toxin receptors (Lenarčič *et al.* 2017), contain primarily very-long-chain fatty acyl moieties (VLCFA) with 24 and 26 carbons in length (Grison *et al.* 2015, Liu *et al.* 2020). Lipid analysis of the *Arabidopsis* PM further revealed that free sterols constitute the most abundant sterol class with sitosterol as the main core structure; phosphatidylcholine (PC) and phosphatidylethanolamine (PE) are the major phospholipid classes with monounsaturated lipids (carrying a single unsaturated fatty acyl moiety per lipid species) as the predominant molecular species (Grison *et al.* 2015, Uemura *et al.* 1995).

Freezing is one of the most frequent environmental stresses that plants encounter. Therefore, plants have evolved a strategy, named cold acclimation (CA), to increase their freezing tolerance already at low temperatures before actual freezing occurs. It has been demonstrated that the proportions of both sphingolipids and sterols decrease under CA condition, whereas phospholipids rise (Uemura *et al.* 1995). In addition, analysis of the lipid species profile has indicated that lipid species with higher unsaturation degree increase proportionally under CA condition, which enhances the fluidity of the PM and thus the freezing tolerance of the cell (Palta *et al.* 1993, Takahashi *et al.* 2016, Uemura *et al.* 1995, Uemura and Steponkus 1994). In our previous work, we developed a wide-ranging LC-MS-based lipidomics platform to profile the lipids of *Arabidopsis* wild-type leaves, and used this to investigate changes in the molecular species of 23 different lipid classes in response to cold and drought stresses (Tarazona *et al.* 2015). The results demonstrated the importance of elucidating lipid classes as well as individual molecular species instead of lipid apparent species only. Here, we employed an in-depth profiling of individual lipid species within all lipid classes present in *Arabidopsis* PM, including additionally the functional lipid classes such as series of GIPCs and phosphoinositides, to understand the molecular modulation of plant PM under environmental stresses.

The physical interactions between complex sphingolipids and sterols are critical for establishing lipid rafts. They may be mediated by the hydroxyl group at the C2 position (α-hydroxylation) on the fatty acid moiety or the hydroxyl group at the C3 or C4 position of the LCB moiety of the Cer backbone via strong hydrogen bonding with free hydroxyl groups at the A ring of sterols (Brown and London 1998, Mamode Cassim *et al.* 2019, Simons and Ikonen 1997). Two fatty acid hydroxylases, *At*FAH1 and *At*FAH2, are capable of introducing an α-hydroxyl group on the fatty acid moiety of complex sphingolipids (Nagano *et al.* 2009), and analysis of their substrate specificity revealed that they selectively hydroxylate either VLCFAs or palmitic acid, respectively (Nagano *et al.* 2012). The double mutant *fah1 fah2* (with T-DNA insertions in the promoter region and the 5th exon, respectively) displays a reduction of α-hydroxylated complex sphingolipids, in combination with a disordered PM structure, elevated salicylic acid (SA) levels and increased disease resistance against biotrophic pathogens (König *et al.* 2012, Lenarčič *et al.* 2017). This further emphasizes the importance of complex sphingolipids in plant cells, despite being minor components in biological membranes.

According to studies on the mammalian systems, lipids are distributed asymmetrically across the lipid bilayer of the PM, with PC and sphingomyelin present mostly within the exoplasmic leaflet whereas PS, PE and PI are within the cytoplasmic leaflet (Yamaji-Hasegawa and Tsujimoto 2006). The plant PM was presumed to possess a similar lipid distribution as in mammalian cells. However, it has been demonstrated that phospholipids are dispersed symmetrically within the PM bilayer of mung beans (Takeda and Kasamo 2001), whereas glucosylceramides (GlcCers) are located within the apoplastic leaflet of PM isolated from summer quash (Lynch and Phinney 1995) and oat roots (Tjellström *et al.* 2010), but to a different extent (98 % and 70 %, respectively). A recent study of the human erythrocytes demonstrated that the asymmetrical distribution occurs at the lipid species level, which impacts on fluidity and membrane organization of each PM monolayer (Lorent *et al.* 2020). However, the transversal lipid distribution of the plant PM is still poorly understood, due to the lack of knowledge on its overall lipid composition, molecular species profile and the precise topology and distribution of individual lipid species across the membrane bilayer.

Building on our previous work (Tarazona *et al.* 2015), we expanded the detection coverage of the lipidomics workflow and implemented it in combination with a subcellular organelle fractionation, to comparatively analyze the lipidomes of PM fractions isolated from *Arabidopsis* WT and the *fah1 fah2* double mutant. Here, we observed that the loss of α-hydoxylases in *Arabidopsis* plants led to similar responses as observed under cold stress at the lipid class level. Nevertheless, elevated levels of sphingolipids, sterols and polyunsaturated glycerolipids in response to CA conditions were observed in both WT and *fah1 fah2* PM preparations, suggesting that the cold-induced responses in *fah1 fah2* are comparable with WT at the molecular lipid species level. The investigation of the transversal lipid distribution revealed that PS resides predominantly within the cytoplasmic leaflet, while PE, PC and PI are mostly equally distributed within both monolayers; higher proportions of GlcCer are found on the apoplastic side; and steryl glycosides (SGs) and acylated steryl glycosides (ASGs) are allocated to the two leaflets in a species-dependent manner.

Results

To understand the function of α-hydroxylated sphingolipids on the organization of the plant PM, we applied a comprehensive lipidomics analysis on PMs purified from rosettes of *Arabidopsis* WT and the sphingolipid fatty acid α-hydroxylase mutant, *fah1 fah2*, grown under non-acclimated and cold-acclimated conditions. Furthermore, the transversal distribution of the most abundant PM-localized phospholipids and glycolipids was determined at the lipid species level.

Enrichment assessment of the plasma membrane fraction by proteomics

PMs were purified from other subcellular membranes via two-phase partitioning (Larsson *et al.* 1994). Furthermore, microsomes (MC) and intracellular membranes (IM), which present the crude membrane fraction and the PM-excluding fraction were generated along this procedure. To ensure and evaluate the purity of the PM isolated from *Arabidopsis* rosettes in correlation with the other fractions including MC, IM and leaf total extract (TE), a wide-coverage shotgun proteomics analysis was performed instead of the conventional immunoblotting, which detects only a selected pool of organelle-specific markers. The identified proteins were filtered based on two criteria: At least two peptide-spectrum matches as well as a threshold of high-confident identification were required for each hit. The resulting list contained 106 proteins identified in the PM, 210 in IM, 534 in MC and 530 in TE. They were subjected to subsequent label-free quantification (LFQ), which compares the protein intensities between samples of different matrixes (Tab. S1). Based on their abundance estimated by the LFQ algorithm, the enrichment of PM-localized proteins in the PM fractions was evaluated. The hierarchical analysis of the subjected proteins displayed that the PM contains a distinct protein profile that is significantly different to those of TE, MC and IM fractions (Fig. 1a). Proteins enriched in the PM fractions with positive LFQ values (Fig. 1b) include typical PM markers such as aquaporins, proton pump ATPases, remorins and syntaxins (Bhat and Panstruga 2005, Marmagne *et al.* 2007, Nühse *et al.* 2003, Uemura *et al.* 2004). The protein profile of our PM correlates well with the PM proteome from *Arabidopsis* cell cultures and leaves (Alexandersson *et al.* 2004, Marmagne *et al.* 2007, Marmagne *et al.* 2004). Furthermore, subcellular localization of the identified membrane-associated or embedded proteins demonstrated that the purity of our PMs reaches 89 % (Fig. 1c), suggesting the here isolated PM fraction is of high purity. The membrane proteins associated with endomembranes (ER, Golgi apparatus and vacuole) and plastids account for 6 % and 4 %, respectively. Noteworthy, typical protein markers, that are commonly used in the immunoblotting approach, targeting plastids, mitochondria and nucleus such as photosystem light harvesting complexes, mitochondrial ATP synthases and ribosomal proteins, respectively, were not identified or strongly reduced in our PM fractions (Tab. S1). This result confirmed the quality of the PM-enriched fractions, demonstrating that major contaminants from other subcellular compartments were eliminated.

Distinct lipids are enriched in the plasma membrane fraction of *Arabidopsis* leaves

To depict the lipid landscape of the PM from *Arabidopsis* WT leaves, we applied a combinatorial approach that incorporates quantitative data for each lipid class of all fractions as determined by TLC-GC, as well as the detailed information on the molecular species profiles obtained via LC-MS/MS profiling from TE versus PM preparations. The TLC-GC analysis of the glycerolipid composition demonstrated that phosphatidylethanolamine (PE; 38.2 ± 12.0 %) is the most abundant glycerolipid in the PM and highly enriched in comparison to the TE, MC and IM fractions (Fig. 2). Collectively, the mixture of phosphatidylserine (PS), phosphatidylinositol (PI) and sulfoquinovosyldiacylglycerol (SQDG) accounts for the most abundant glycerolipids in MC (47.9 ± 4.6 %); nevertheless, it constitutes a considerable amount in the PM (26.5 ± 7.1 %) as well.

It should be noted that due to the limitations of the chromatographic separation, these three lipid classes were closely resolved on the TLC plate, that it was not feasible to quantify them individually. Phosphatidylcholine (PC; 6.5 ± 6.4 %) and phosphatidylglycerol (PG; 13.8 ± 4.0 %) are only minor phospholipids of the PM. Low amounts of the glyceroglycolipids monogalactosyldiacylglycerol (MGDG; 5.4 ± 1.0 %) and digalactosyldiacylglycerol (DGDG; 9.6 ± 4.9 %) were detected in the PM, whereas they represent the most abundant lipid classes in the TE (60.3 ± 5.0 %) and the IM (55.1 %± 10.9 %). Further determinations of the absolute abundance of sphingolipids and sterols are in progress.

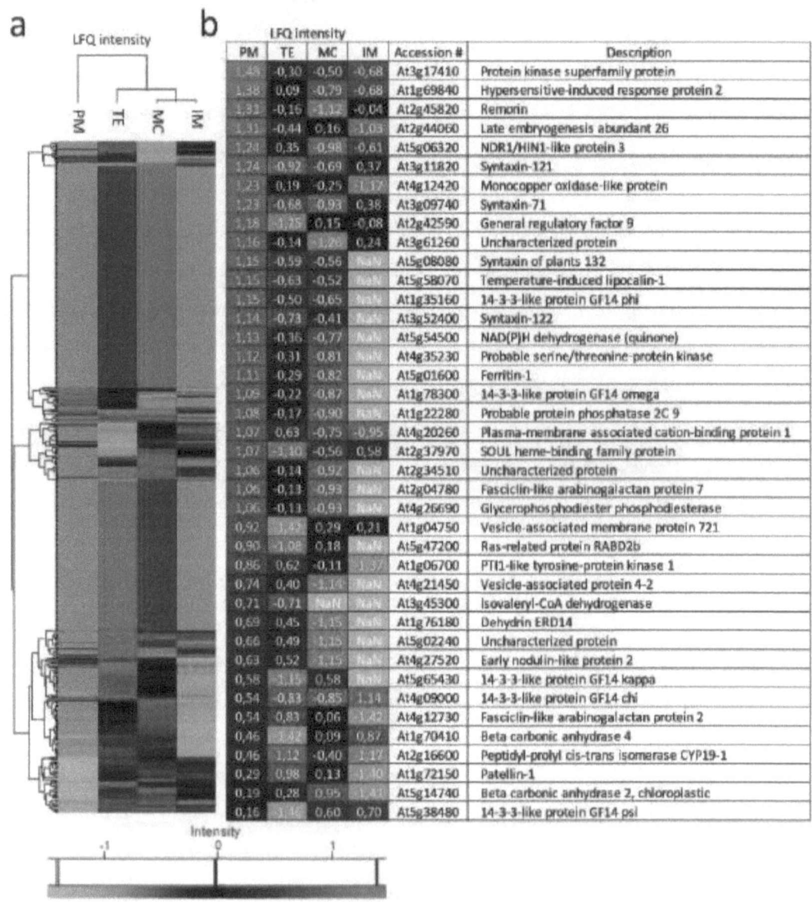

LFQ intensity

PM	TE	MC	IM	Accession #	Description
1,48	-0,30	-0,50	-0,68	At3g17410	Protein kinase superfamily protein
1,38	0,09	-0,79	-0,68	At1g69840	Hypersensitive-induced response protein 2
1,31	-0,16	-1,12	-0,04	At2g45820	Remorin
1,31	-0,44	0,16	-1,03	At2g44060	Late embryogenesis abundant 26
1,24	0,35	-0,98	-0,61	At5g06320	NDR1/HIN1-like protein 3
1,24	0,92	-0,69	0,37	At3g11820	Syntaxin-121
1,23	0,19	-0,25	-1,17	At4g12420	Monocopper oxidase-like protein
1,23	-0,68	0,93	0,38	At3g09740	Syntaxin-71
1,18	-1,25	0,15	-0,08	At2g42590	General regulatory factor 9
1,16	-0,14	-1,26	0,24	At3g61260	Uncharacterized protein
1,15	-0,59	-0,56	NaN	At5g08080	Syntaxin of plants 132
1,15	-0,63	-0,52	NaN	At5g58070	Temperature-induced lipocalin-1
1,15	-0,50	-0,65	NaN	At1g35160	14-3-3-like protein GF14 phi
1,14	-0,73	-0,41	NaN	At3g52400	Syntaxin-122
1,13	-0,36	-0,77	NaN	At5g54500	NAD(P)H dehydrogenase (quinone)
1,12	0,31	0,81	NaN	At4g35230	Probable serine/threonine-protein kinase
1,11	0,29	0,82	NaN	At5g01600	Ferritin-1
1,09	-0,22	-0,87	NaN	At1g78300	14-3-3-like protein GF14 omega
1,08	-0,17	-0,90	NaN	At1g22280	Probable protein phosphatase 2C 9
1,07	0,63	-0,75	-0,95	At4g20260	Plasma-membrane associated cation-binding protein 1
1,07	-1,10	-0,56	0,58	At2g37970	SOUL heme-binding family protein
1,06	-0,14	0,92	NaN	At2g34510	Uncharacterized protein
1,06	-0,13	0,93	NaN	At2g04780	Fasciclin-like arabinogalactan protein 7
1,06	-0,13	-0,93	NaN	At4g26690	Glycerophosphodiester phosphodiesterase
0,92	1,42	0,29	0,21	At1g04750	Vesicle-associated membrane protein 721
0,90	-1,08	0,18	NaN	At5g47200	Ras-related protein RABD2b
0,86	0,62	-0,11	-1,37	At1g06700	PTI1-like tyrosine-protein kinase 1
0,74	0,40	-1,14	NaN	At4g21450	Vesicle-associated protein 4-2
0,71	-0,71	NaN	NaN	At3g45300	Isovaleryl-CoA dehydrogenase
0,69	0,45	-1,15	NaN	At1g76180	Dehydrin ERD14
0,66	0,49	1,15	NaN	At5g02240	Uncharacterized protein
0,63	0,52	1,15	NaN	At4g27520	Early nodulin-like protein 2
0,58	1,15	0,58	NaN	At5g65430	14-3-3-like protein GF14 kappa
0,54	-0,83	-0,85	1,14	At4g09000	14-3-3-like protein GF14 chi
0,54	0,83	0,06	-1,42	At4g12730	Fasciclin-like arabinogalactan protein 2
0,46	1,42	0,09	0,87	At1g70410	Beta carbonic anhydrase 4
0,46	1,12	-0,40	1,17	At2g16600	Peptidyl-prolyl cis-trans isomerase CYP19-1
0,29	0,98	0,13	-1,40	At1g72150	Patellin-1
0,19	0,28	0,95	-1,41	At5g14740	Beta carbonic anhydrase 2, chloroplastic
0,16	1,16	0,60	0,70	At5g38480	14-3-3-like protein GF14 psi

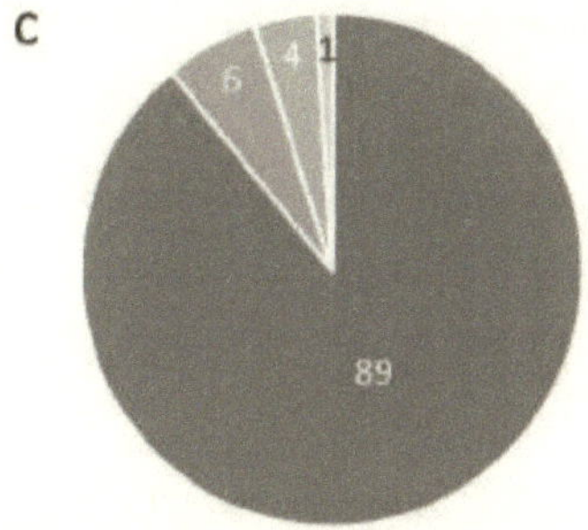

Figure 1. Label-free quantification of the identified proteins by shotgun proteomics.

Proteins extracted from different membrane fractions were analyzed via shotgun proteomics and quantified by label-free quantification (LFQ). (a) Hierarchical analysis illustrates the abundance of pooled identified proteins from all membrane fractions. (b) Proteins with positive LFQ intensities in the PM fractions were extracted and listed with their accession numbers and descriptions. High to low intensities are indicated in the red to green gradient. Data represent mean LFQ values from three independent experiments. (c) Intensities of the membrane-associated and embedded proteins assigned to subcellular membranes were summed-up and visualized by the pie cart. Blue: plasma membranes; orange: endomembranes (ER, Golgi apparatus and vacuole); gray: plastids; yellow: others; numbers in %. IM, intracellular membranes; MC, microsomes; NaN, not a number; PM, plasma membrane; TE, total extract.

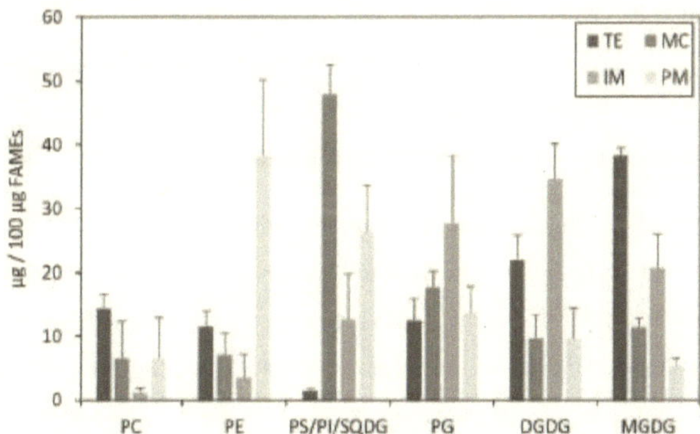

Figure 2. Glycerolipid class profiles of all membrane fractions isolated from *Arabidopsis* leaves.
Glycerolipids of total extract (TE), microsomes (MC), intracellular membranes (IM) and plasma membrane (PM) and were analyzed quantitatively by TLC-GC approach. Data represent mean values in µg from three independent experiments ± SD. To normalize the loading lipid amounts between different membrane fractions, aliquots of lipid extracts that contain 100 µg fatty acid methyl esters (FAMEs) were loaded onto the TLC plates. DGDG, digalactosyldiacylglycerol; MGDG, monogalactosyldiacylglycerol; PC, phosphatidylcholine; PE, phosphatidylethanolamine; PG, phosphatidylglycerol; PI, phosphatidylinositol; PS, phosphatidylserine; SQDG, sulfoquinovosyldiacylglycerol.

Lipidomics analysis reveal the molecular lipid species composition of the plasma membrane fraction

PMs isolated from WT plants grown under optimal non-acclimated (NA) condition were investigated by LC-MS-based lipidomics analysis to assess the prevailing molecular species profiles of glycerolipids, sphingolipids, and sterols.

Sphingolipids are mostly found in extraplastidial membranes. Profiling the molecular composition of sphingolipids revealed that hydroxysphinganine (18:1;3) is the predominant constituent of both LCBs (57.2 ± 10.3 % of total LCBs) and phosphorylated LCBs (LCB-P; 70.2 ± 1.8 %) (Fig. 3, Tab. S2). On the other hand, complex sphingolipids comprise a more diverse selection regarding their LCB backbones including 18:1;3, sphingenine (18:1;2) and sphingadiene (18:2;2) (Fig. 4). Higher signals were obtained from the hydroxylated complex sphingolipids, which are acylated with α-hydroxy fatty acyl moieties. In WT PM, hydroxylated ceramide (hCer) with 18:1;3/24:0 (17.6 ± 4.3 % of total Cers including both non- and hydroxylated variants) and 18:1;3/24:1 (19.7 ± 3.3 %) are the most prominent Cer species. Hydroxylated glucosylceramide (hGlcCer) with 18:1;3/16:0 (23.0 ± 4.1 % of total GlcCers including both non- and hydroxylated variants), 18:1;3/22:0 (10.3 ± 0.6 %) and 18:1;3/24:1 (44.8 ± 3.9 %) are the major GlcCer species. The sequential extension of the phosphoinositol head group of the Cers, with one, two and three hexoses and/or hexose derivatives generates series 0, A and B GIPCs, respectively. Noteworthy, series A and B GIPCs are enriched in WT PM with higher proportion contributed by series A GIPCs. The major GIPC species in WT PM are the hydroxylated hexose-carrying GIPC (hH-GIPC) containing 18:1;3/24:0 (37.2 ± 11.2 % of total GIPCs including bot non- and hydroxylated variants) and 18:1;3/24:1 (20.7 ± 5.9 %) (Fig. 5).

Sterols are additional critical components in determining membrane fluidity and organization. Most of the sterol species found in both PM and TE are sitosterol and campesterol as well as their esterified derivatives (Fig. 6). Similar species profiles of sterols were detected in all extracts, except the free sitosterols and free isofucosterols are enriched whereas the free campesterols and the free cholesterols are reduced in TE when compared to PM. In WT PM, species with sitosterol and campesterol backbones collectively account for 76.8 % of free sterols (FSs), 90.5 % of steryl glycosides (SGs), 88.2 % of steryl esters (SEs) and 90.7 % of acylated steryl glycosides (ASGs). In addition, sitosteryl esters containing 18:2 and 18:3 acyl moieties are the predominant SE species (43.3 ± 27.8 % and 32.3 ± 20.4 %, respectively) and acylated sitosteryl glycoside containing 16:0 is the most prominent ASG species (35.5 ± 14.9 %).

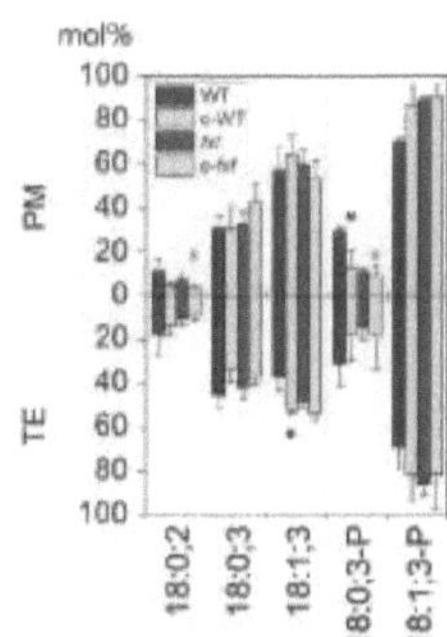

Figure 3. LCB and LCB-P compositions of PM and TE isolated from *Arabidopsis* WT and *fah1 fah2* plants under non- and cold-acclimated conditions.

LCBs and LCB-Ps of isolated plasma membrane (PM) and total extract (TE) fractions from wild-type and *fah1 fah2 Arabidopsis* grown under normal (WT and *fxf*) and cold-acclimated (CA) conditions (c-WT and c-*fxf*) were analyzed by LC-MS/MS. Data represent mean values of mol % of an individual lipid species in the according lipid class from three independent experiments ± SD. Black and blue asterisks indicate significant differences between growth conditions in WT and *fah1 fah2* background, respectively (*P < 0.05; Student's t-test). LCB, long-chain base; LCB-P, phosphorylated long-chain base.

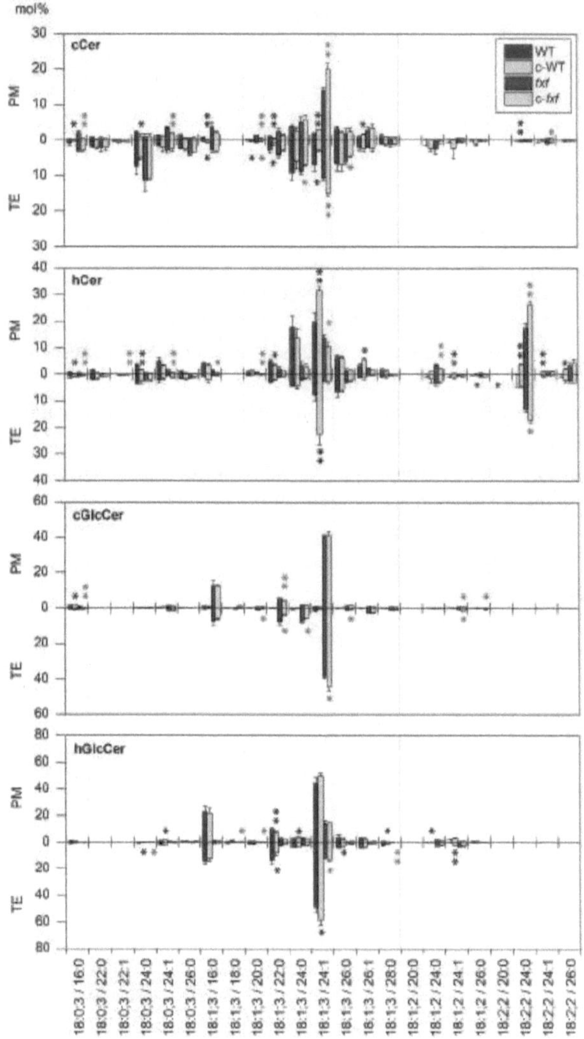

Figure 4. Cer and GlcCer compositions of PM and TE isolated from *Arabidopsis* WT and *fah1 fah2* plants under non- and cold-acclimated conditions.

Cers and GlcCers of isolated plasma membrane (PM) and total extract (TE) fractions from wild-type and *fah1 fah2 Arabidopsis* grown under normal (WT and *fxf*) and cold-acclimated (CA) conditions (c-WT and c-*fxf*) were analyzed by LC-MS/MS. Data represent mean values of mol % of an individual lipid species in the total Cers and GlcCers, respectively, from three independent experiments ± SD. Non-hydroxylated and hydroxylated species from ceramides (cCer and hCer) and glucosylceramides (cGlcCer and hGlcCer) are illustrated separately. Black and blue asterisks indicate significant differences between growth conditions in WT and *fah1 fah2* background, respectively (*P < 0.05, ** P < 0.01; Student's t-test). Cer, ceramide; GlcCer, glucosylceramide.

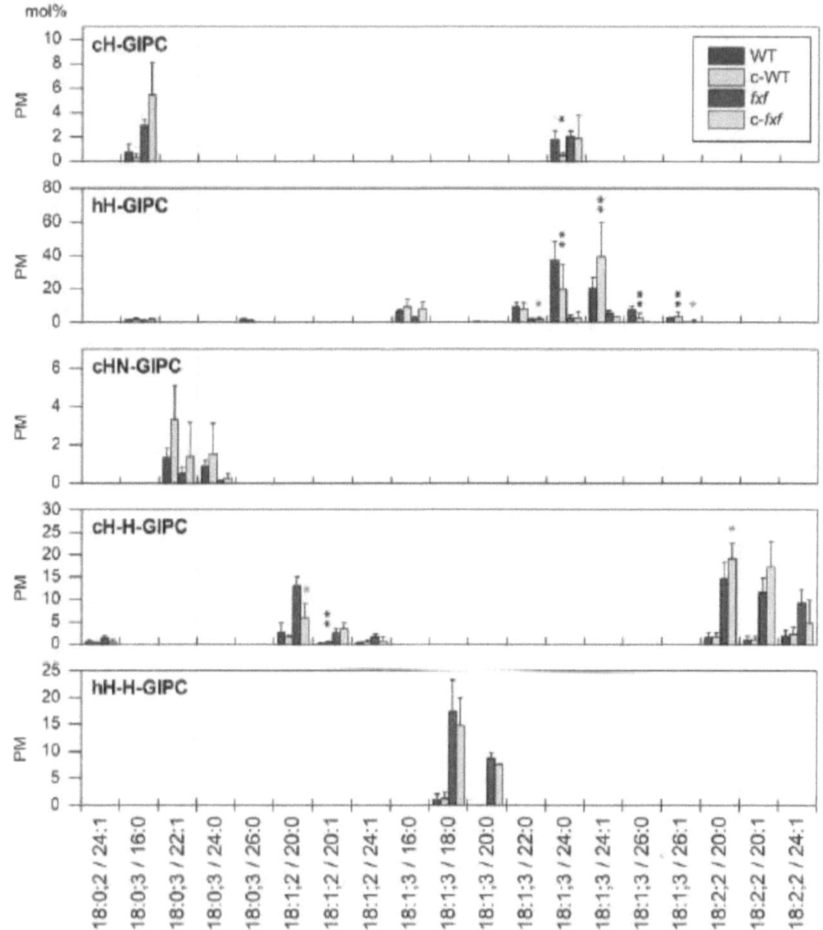

Figure 5. GIPC compositions of PM isolated from *Arabidopsis* WT and *fah1 fah2* plants under non- and cold-acclimated conditions.

Series A and B GIPCs of isolated PM fractions from wild-type and *fah1 fah2 Arabidopsis* grown under normal (WT and *fxf*) and cold-acclimated (CA) conditions (c-WT and c-*fxf*) were analyzed by LC-MS/MS. Data represent mean values of mol % of an individual lipid species in total GIPCs from three independent experiments ± SD. Non-hydroxylated and hydroxylated species (cGIPC and hGIPC) of series A GIPCs (H-GIPC and HN-GIPC) and series B GIPCs (H-H-GIPC) are illustrated separately. Black and blue asterisks indicate significant differences between growth conditions in WT and *fah1 fah2* background, respectively (*P < 0.05, ** P < 0.01; Student's t-test). GIPC, glycosyl inositol phosphoceramide; H, hexosyl; HN, *N*-acetylhexosaminyl, H-H, dihexosyl.

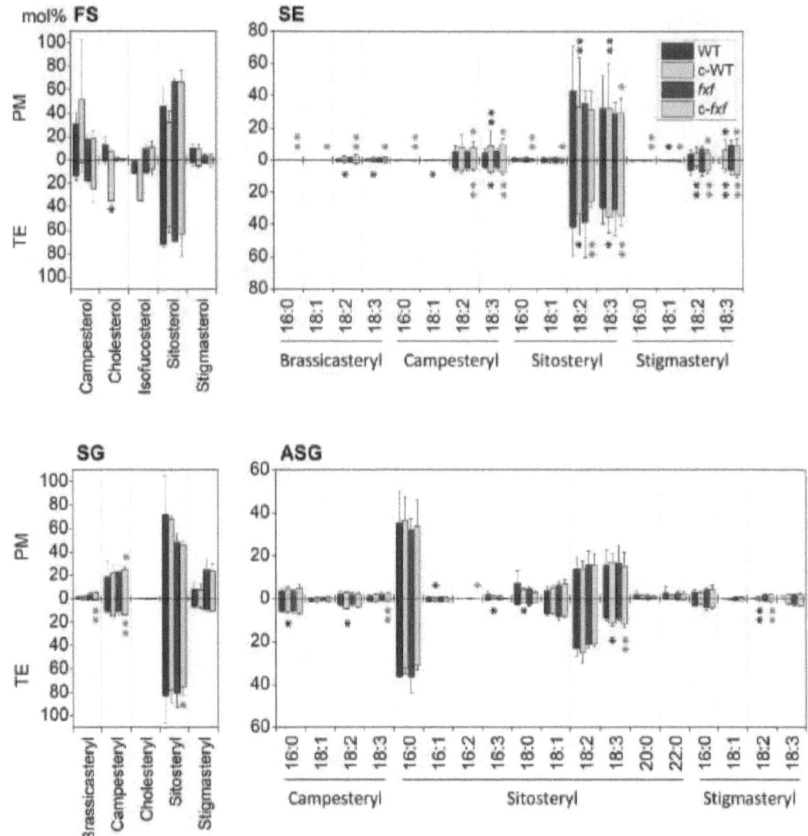

Figure 6. Sterol compositions of PM and TE isolated from *Arabidopsis* WT and *fah1 fah2* plants under non- and cold-acclimated conditions.

Sterols of isolated plasma membrane (PM) and total extract (TE) fractions from wild-type and *fah1 fah2 Arabidopsis* grown under normal (WT and *fxf*) and cold-acclimated (CA) conditions (c-WT and c-*fxf*) were analyzed by LC-MS/MS. Data represent mean values of mol % of an individual lipid species in the according lipid class from three independent experiments ± SD. Black and blue asterisks indicate significant differences between growth conditions in WT and *fah1 fah2* background, respectively (*P < 0.05, ** P < 0.01; Student's t-test). ASG, acyl steryl glycoside; FS, free sterol; SE, sterol ester; SG, steryl glycoside.

The molecular species profiles of glycerophospholipids revealed that the monounsaturated lipid species containing a single unsaturated fatty acyl moiety per lipid species, are the predominant types in WT PM (Fig. 7). Noteworthy, the 16:0-containing monounsaturated lipids including 16:0/18:2 and 16:0/18:3 constitute major proportions of PCs (51.4 ± 8.9 % of total PCs), PEs

(67.8 ± 4.2 %), PSs (40.6 ± 5.5 %), PIs (72.9 ± 16.6 %), phosphatidylinositol monophosphates (PIPs; 100 ± 46.1 %) and phosphatidylinositol bisphosphates (PIP$_2$s; 72.9 ± 12.0 %). Diunsaturated lipid species, containing two unsaturated fatty acyl moieties per molecule, account for much lower percentages in the mentioned lipid classes in PMs, although they feature comparable levels as the monounsaturated species in PCs and PEs in TEs of WT (Fig. 7). Furthermore, unique PE and PS varieties, which carry VLCFA moieties (22, 24 and 26 carbons), accumulate in the PM. For instance, PS (18:2/26:0) and PS (18:3/26:0) are more than 5-fold enriched in the PM in comparison to the TE from WT. It should be noted that the occurrence of PIPs and PIP$_2$s, of which the major molecular species can be detected with the standardized extraction procedure by our high-throughput LC-MS system, were specific to the PM extracts. In contrast to the other glycerophospholipid classes, high levels of diunsaturated species (18:2/18:2, 18:2/18:3 and 18:3/18:3; 38.5 ± 2.4 %) were maintained for phosphatidic acids (PAs) in the PMs, although respective 16:0-containing monounsaturated lipid species still represent the predominant PA fraction (16:0/18:2 and 16:0/18:3; 50.2 ± 3.6 %). Noteworthy, considerable differences between the PG profiles of PM and TE were revealed; for instance, PG (16:1/18:3), which is a characteristic PG species of plastidial origin, is a minor component of the PM (1.8 ± 0.7 %), but abundant in the TE (29.6 ± 2.1 %). In contrast, disaturated PG species, containing two saturated fatty acyl moieties per lipid species, such as PG (16:0/16:0; 27.3 ± 9.7 %), are extremely enriched in the PM preparation as opposed to the TE ones. Otherwise, disaturated lipids form only a very small part of the glycerophospholipid classes in the PM. As already observed by TLC, a small portion of glyceroglycolipids can be also detected in the PM. While MGDG (16:3/18:3) and DGDG (18:3/18:3) are the most abundant species in both PM and TE, a considerable enrichment of 16:0-containing monounsaturated glyceroglycolipids was detected in the PM, indicative of the transport of eukaryotic glyceroglycolipid moieties from the chloroplast to the PM (Fig. 8). Taken together, the distinct PG, MGDG and DGDG profiles of the PM and the TE indicate that the PM fractions have minimum lipid contaminants from plastidial membranes. Corresponding to glycerophospholipids, the 16:0-containing monounsaturated lipids including 16:0/18:2 and 16:0/18:3 also constitute major proportions of diacylglycerols (DAGs; 67.4 ± 5.2 %). The predominant species of the triacylglycerols (TAGs) in the PMs including 52:4, 52:5 and 52:6 (36.4 ± 11.8 %), presumably comprise 16:0, 18:2 and 18:3 acyl moieties (Fig. 9). The sulfur-containing lipids, sulfoquinovosyldiacylglycerols (SQDGs) are only present in the WT TE in which SQDG (34:3) is the most abundant species (Fig. S1). In summary, the in-depth lipidomics analyses revealed that sphingolipids with α-hydroxylated fatty acid moieties, sterols with sitosterols and campesterols as the core structures and glycerophospholipids with an unsaturated fatty acyl moiety per lipid species are the predominant ones in WT PM. Only specific PG, MGDG and DGDG species are enriched in the PMs, again indicating that our PMs are of high purity.

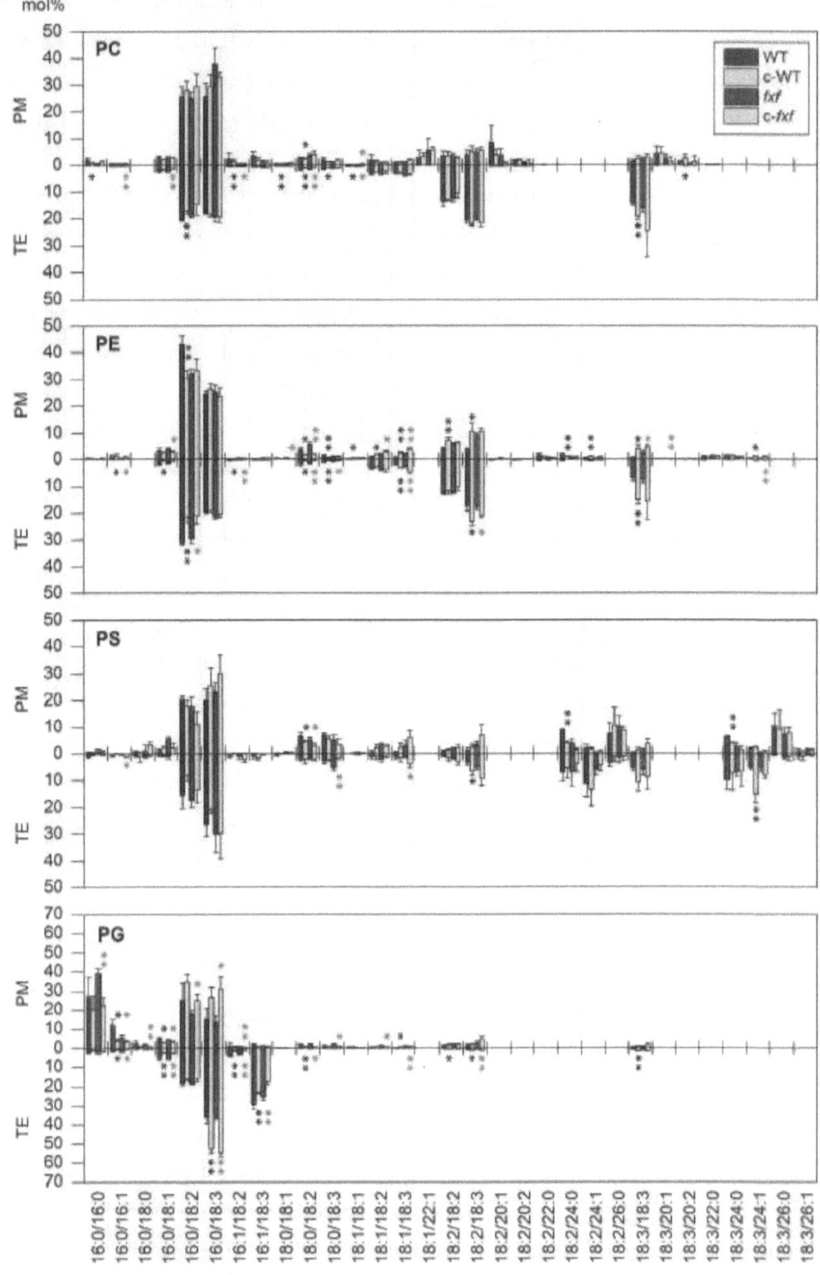

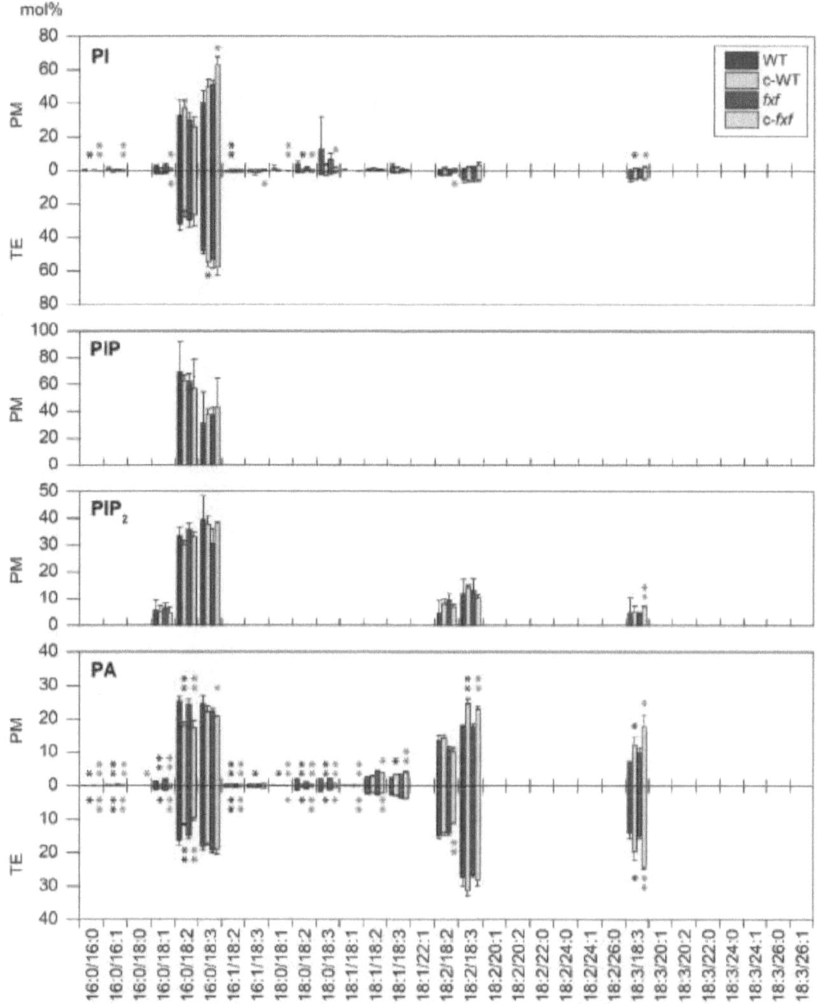

Figure 7. Phospholipid composition of PM and TE isolated from *Arabidopsis* WT and *fah1 fah2* plants under non- and cold-acclimated conditions.

Glycerophospholipids of isolated plasma membrane (PM) and total extract (TE) fractions from wild-type and *fah1 fah2 Arabidopsis* grown under normal (WT and *fxf*) and cold-acclimated (CA) conditions (c-WT and c-*fxf*) were analyzed by LC-MS/MS. Data represent mean values of mol % of an individual lipid species in the according lipid class from three independent experiments ± SD. Black and blue asterisks indicate significant differences between growth conditions in WT and *fah1 fah2* background, respectively (*P < 0.05, ** P < 0.01; Student's t-test). PA, phosphatidic acid; PC, phosphatidylcholine; PE, phosphatidylethanolamine; PI, phosphatidylinositol; PIP and PIP$_2$, phosphatidylinositol mono- and bisphosphates; PS, phosphatidylserine; PG, phosphatidylglycerol.

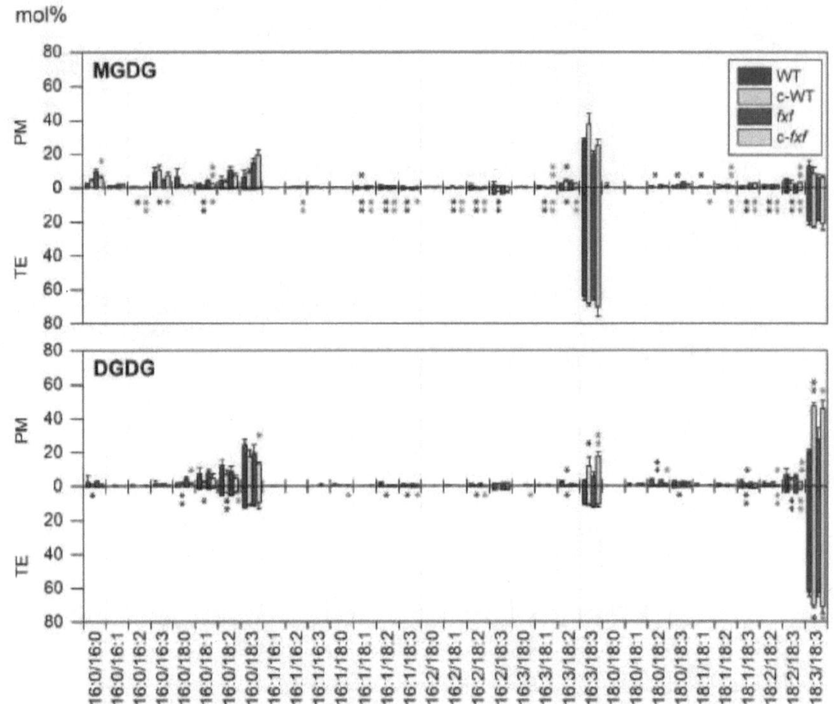

Figure 8. Glyceroglycolipid composition of PM and TE isolated from *Arabidopsis* WT and *fah1 fah2* plants under non- and cold-acclimated conditions.

Glyceroglycolipids of isolated plasma membrane (PM) and total extract (TE) fractions from wild-type and *fah1 fah2* Arabidopsis grown under normal (WT and *fxf*) and cold-acclimated (CA) conditions (c-WT and c-*fxf*) were analyzed by LC-MS/MS. Data represent mean values of mol % of an individual lipid species in the according lipid class from three independent experiments ± SD. Black and blue asterisks indicate significant differences between growth conditions in WT and *fah1 fah2* background, respectively (*P < 0.05, ** P < 0.01; Student's t-test). DGDG, digalactosyldiacylglycerol; MGDG, monogalactosyldiacylglycerol.

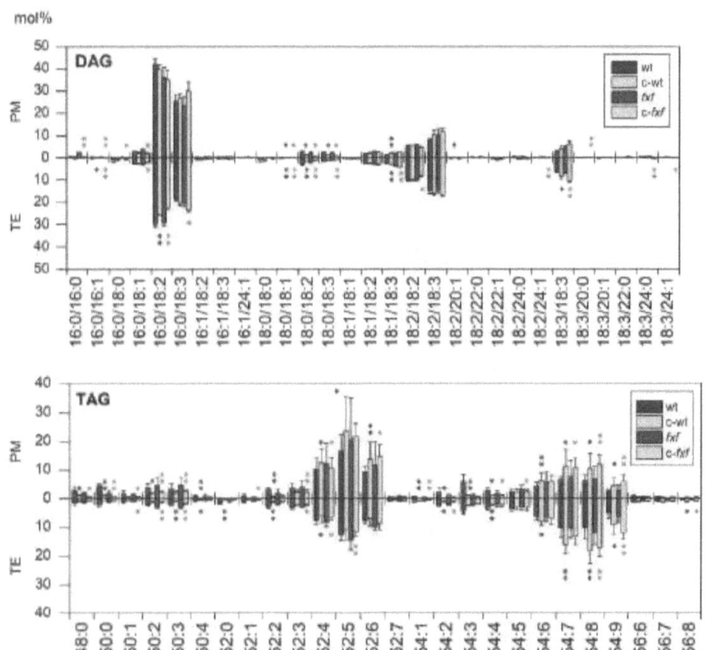

Figure 9. Neutral glycerolipid composition of PM and TE isolated from *Arabidopsis* WT and *fah1 fah2* plants under non- and cold-acclimated conditions.

Neutral glycerolipids of isolated plasma membrane (PM) and total extract (TE) fractions from wild-type and *fah1 fah2* Arabidopsis grown under normal (WT and *fxf*) and cold-acclimated (CA) conditions (c-WT and c-*fxf*) were analyzed by LC-MS/MS. Data represent mean values of mol % of an individual lipid species in the according lipid class from three independent experiments ± SD. Black and blue asterisks indicate significant differences between growth conditions in WT and *fah1 fah2* background, respectively (*P < 0.05, ** P < 0.01; Student's t-test). DAG, diacylglycerol; TAG, triacylglycerol.

Loss of sphingolipid α-hydroxylase activity leads to reduced levels of complex sphingolipids and sterols

In order to understand the effects of α-hydroxylated sphingolipids on the PM organization, individual molecular species from both TE and PM preparations of *fah1 fah2* mutant plants were equally profiled in comparison to WT (Figs 3-9). All detected species per lipid class were added, the relative percentage of each lipid class of both PM and TE preparations determined, and finally the fold changes between *fah1 fah2* and WT plants calculated (Fig. 10, Tab. S3). In general, the observed fold changes between the WT and the mutant plants differ significantly between the TE and the PM preparations, supporting PM-specific changes within the lipid composition of *fah1 fah2* mutant plants. The lipid profile of the *fah1 fah2* TE is characterized by a strong loss of GlcCer and H-GIPC and mild reductions of PC, PE and PS when compared to WT (Fig. 10). On the other hand, high amounts of Cer and moderate amounts of LCB, LCB-P, SE, SG, ASG, DAG and TAG accumulate in *fah1 fah2* TE. This is corroborated by previous studies, indicating an elevated amount of LCB and reduced amounts of GlcCer and H-GIPC in TEs of *fah1 fah2* (König *et al.* 2012, Lenarčič *et al.* 2017). Whereas GlcCer, H-GIPC, N-acetylhexosaminyl (HN)-GIPC, SG, ASG, PC, PE, PS, DAG and TAG are strongly reduced in *fah1 fah2* PMs; LCB-P, dihexose-carrying series B GIPC (H-H-GIPC), SE and PA are highly increased in comparison to PMs isolated from WT. The elevated amounts of these lipid classes in the *fah1 fah2* PM possibly suggest a role in rescuing the disturbed membrane composition caused by the loss of α-hydroxyl groups in the fatty acyl moieties. Noteworthy, some lipid classes such as DAG, TAG, Cer, HN-GIPC, SG and ASG show great reduction in *fah1 fah2* PM, but a mild enrichment in *fah1 fah2* TE when compared to the respective WT profiles. Again, this differential enrichment and distribution of the lipid classes indicate that the PM possesses a specific lipid profile in comparison to other intracellular membranes.

The effects of α-hydroxylase mutations on the individual molecular species level within the PM were further investigated. Similar species profiles of LCBs were detected in all extracts, whereas higher levels of LCB-P (18:1;3) were observed in both *fah1 fah2* PM and *fah1 fah2* TE extracts (1.3 and 1.2-fold increase, respectively) (Fig. 3). It should be noted that complex sphingolipids carrying α-hydroxylated fatty acyl moieties were still detected, which derive from either residual FAH1 activity or other yet unknown FAH proteins (König *et al.* 2012). Nevertheless, a strong reduction of the hydroxylated complex sphingolipids is characteristic for *fah1 fah2* plants and especially prominent in the profile of GlcCer molecular species (Fig. 4). Significant decrease of the proportion of hCer (18:1;3/24:0) and hCer (18:1;3/24:1), as well as hGlcCer (18:1;3/16:0), hGlcCer (18:1;3/22:0) and hGlcCer (18:1;3/24:1) were observed in *fah1 fah2* PM when compared to WT PM. On the other hand, non-hydroxylated ceramides and glucosylceramides (cCers and cGlcCers, respectively) such as cCer (18:1;3/16:0) and cCer (18:1;3/24:1), as well as cGlcCer (18:1;3/16:0), cGlcCer (18:1;3/22:0) and cGlcCer (18:1;3/24:1) are significantly increased. The changes of the Cer and GlcCer profiles between *fah1 fah2* and WT TEs followed a similar tendency with some exceptions. For instance, cCer (18:1;3/24:1) levels increase strongly in the *fah1 fah2* PM to 7.51-fold of the respective species in the WT PM, but were only mildly enriched in the respective TE extracts. Noteworthy, while most of the hCer species display diminished relative amounts in *fah1 fah2* plants, the proportions of hCer species with 18:2;2 backbones are elevated. Series A and B GIPCs were detectable in the PM fractions where complex sphingolipids are enriched, but not in the corresponding TE samples. Series A GIPC including H-GIPC and HN-GIPC are depleted in *fah1 fah2* PM as opposed to the WT sample (Fig. 10). This depletion resulted majorly from the reduced relative amount of hH-GIPCs, especially from the most abundant species, 18:1;3/24:0 and 18:1;3/24:1 (Fig. 5). On the other hand, non-hydroxylated H-GIPCs (cH-GIPCs) are highly increased while non-hydroxylated HN-GIPCs (cHN-GIPCs) are decreased in the *fah1 fah2* PM as compared to the WT. Noteworthy, all series B GIPCs, including both the non-hydroxylated and hydroxylated species of H-H-GIPCs (hH-H-GIPCs and cH-H-GIPCs, respectively), strongly accumulate in the *fah1 fah2* PMs.

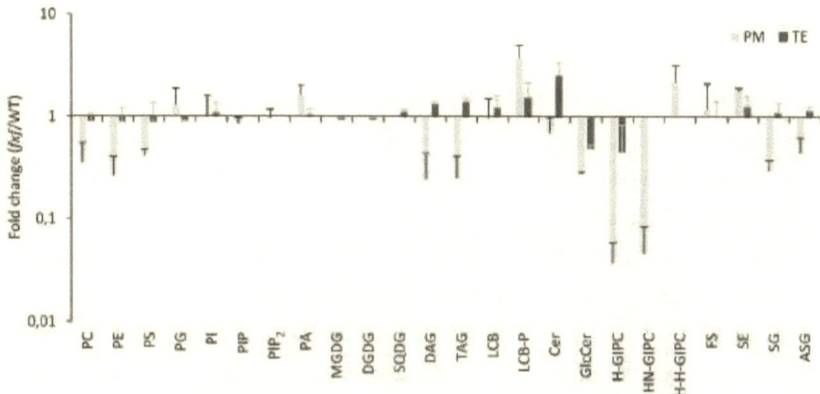

Figure 10. Fold changes between lipid classes of PM and TE from *Arabidopsis* WT and *fah1 fah2* under non-acclimated condition.
Glycerolipids, sphingolipids and sterols of plasma membrane (PM) and total extract (TE) from *Arabidopsis* leaves were analyzed by LC-MS/MS approach. Based on sums of peak areas from all detected species in each lipid class, the relative proportion of each lipid class of PM and TE, and the fold changes between *fah1 fah2* (*fxf*) and wild-type (WT) plants were calculated. Data represent fold changes between *fxf* and WT plants from three independent experiments ± SD. ASG, acyl steryl glycoside; Cer, ceramide; DAG, diacylglycerol; DGDG, digalactosyldiacylglycerol; FS, free sterol; GlcCer, glucosylceramide; H-, HN- and H-H-GIPC, hexosyl-,*N*-acetylhexosaminyl- and dihexosyl-glycosyl inositol phosphoceramide; LCB and LCB-P, long-chain base and its phosphorylated form; MGDG, monogalactosyldiacylglycerol; PA, phosphatidic acid; PC, phosphatidylcholine; PE, phosphatidylethanolamine; PG, phosphatidylglycerol; PI, phosphatidylinositol; PIP and PIP₂, phosphatidylinositol mono- and bisphosphate; PS, phosphatidylserine; SE, steryl ester; SG, steryl glycoside; SQDG, sulfoquinovosyldiacylglycerol; TAG, triacylglycerol.

Together with sphingolipids, sterols and sterol derivatives serve as fundamental components in PM microdomains or lipid rafts (Laloi *et al.* 2007, Lefebvre *et al.* 2007, Schrick *et al.* 2012). Their molecular species in *fah1 fah2* mutant plants were likewise analyzed in TE and PM preparations. In comparison to the WT, SE levels increase while SG and ASG decrease in the *fah1 fah2* PMs, as already observed for the complex sphingolipids (Fig. 10). In addition, although the overall content of FS stays consistent, lower proportions of campesterols, but higher proportions of isofucosterols and sitosterols are present in the *fah1 fah2* PM (Fig. 6). The relative amount of 18:3-stigmasteryl ester and stigmasteryl glycosides are increased, whereas sitosteryl glycosides are reduced in the *fah1 fah2* PM. Interestingly, while the overall content is depleted in the *fah1 fah2* PM, no significant changes were detected with regards to the ASG molecular profile. In general, the overall sterol contents of the *fah1 fah2* TE as well as its molecular species profiles remain steady comparing to the WT TE.

The lipid compositions of the glycerophospholipids in the *fah1 fah2* plants were very similar to those observed in WT, with 16:0-containing monounsaturated varieties still presenting the predominant glycerophospholipid species in *fah1 fah2* plants (Fig. 7). Nevertheless, relative amounts of several glycerophospholipid species increase in the mutant PM; for instance, PE (18:1/18:3), PE (18:2/18:3), PE (18:3/18:3) and PS (18:1/18:2) display a 2.4 to 4.7-fold enrichment as compared to the WT PM. On the other hand, PE (18:2/24:0), PS (18:3/24:1), PS (18:3/24:0), PS (18:2/24:0) and PG (16:0/16:1) display a significant reduction (0.2 to 0.5-fold) in the *fah1 fah2* PM. Only slight increases of 18:1/18:3-containing PCs, PEs and PAs, as well as mild reductions of PC (16:0/18:2) and PC (18:0/18:2) were detected in the *fah1 fah2* TE. Concerning the glyceroglycolipids, the accumulation of 16:0-containing species (1.9-3.5 fold) and concomitant reduction of 18:3-containing species (0.45-0.7 fold) were observed in the MGDG profile, however not in the DGDG profile of the *fah1 fah2* PMs (Fig. 8). In addition, only TAG (50:0) exhibits a 0.39-fold reduction, while the DAG profile of the *fah1 fah2* PMs remains very similar to the one observed in WT (Fig. 9). Altogether, the *fah1 fah2* PMs contain strongly reduced levels of hydroxylated complex sphingolipids (especially GlcCer) and sterols (SG and ASG), but accumulated levels of series B GIPC. On the other hand, the overall species profiles of glycerolipids of the *fah1 fah2* PMs are equivalent to the WT PMs with monounsaturated varieties accounting for the most predominant glycerophospholipid species.

Unsaturation degrees of glycerolipids, sphingolipids and sterols increase under cold acclimation in the wild-type plasma membrane

To address the remodeling response of the PM lipidome upon cold stress, fractionation preparations were also performed on additional *Arabidopsis* WT and *fah1 fah2* mutant plants, grown under CA-conditions. As opposed to the control plants (referred to as NA (non-acclimated) plants and propagated for 31 days under standard growth conditions), 3 weeks old plants were moved to growth chambers at 4 °C for an additional ten days to induce cold acclimation. All detected sphingolipid classes, including LCB, LCB-P, Cer, GlcCer, series A and B GIPCs, display reduced relative amounts under CA conditions as compared to NA conditions (Fig. 11). Nevertheless, the most abundant lipid species of the respective lipid class remain the same in the WT PM preparations (Fig. 3-5). Regarding the complex sphingolipids, the hydroxylated unsaturated species increase while the non-hydroxylated saturated species decrease under CA. For instance, hCer (18:1;3/24:1) and hGlcCer (18:1;3/24:1) highly accumulated while cCer (18:1;3/22:0) and cGlcCer (18:1;3/22:0) are strongly reduced in both PM and TE samples of the WT (Fig. 4). In addition, larger proportions of hH-GIPC (18:1;3/24:1) and hH-GIPC (18:1;3/26:1) were detected in the WT PM (Fig. 5). Interestingly, Cer species with LCB (18:2;2) backbones increase under CA in both WT PM and TE extracts, regardless of the saturation degrees of the acyl moieties (Fig. 4).

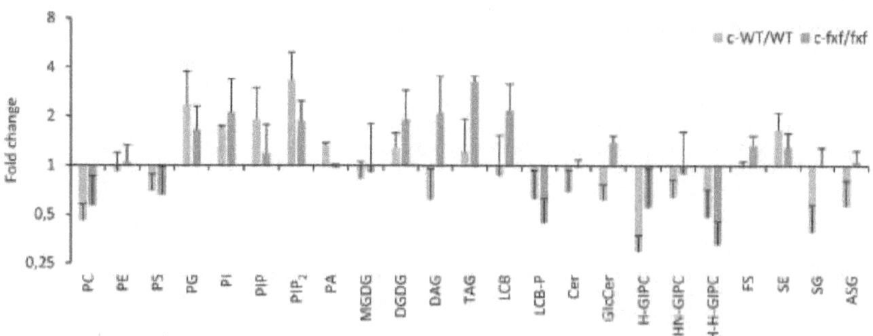

Figure 11. Fold changes between lipid classes of PM from *Arabidopsis* WT and *fah1 fah2* under cold-acclimated condition.

Glycerolipids, sphingolipids and sterols of plasma membrane (PM) from *Arabidopsis* wild type (WT) and *fah1 fah2* leaves grown under cold-acclimated (CA) conditions were analyzed by LC-MS/MS approach. Based on sums of peak areas from all detected species in each lipid class, the relative proportion of each lipid class of *fah1 fah2* (*fxf*) and wild-type (WT) extracts, and the fold changes between cold- and non-acclimated conditions were calculated. Data represent fold change between cold- and non-acclimated plants from three independent experiments ± SD. ASG, acyl steryl glycoside; Cer, ceramide; DAG, diacylglycerol; DGDG, digalactosyldiacylglycerol; FS, free sterol; GlcCer, glucosylceramide; H-, HN- and H-H-GIPC, hexosyl-,*N*-acetylhexosaminyl- and dihexosyl-glycosyl inositol phosphoceramide; LCB and LCB-P, long-chain base and its phosphorylated form; MGDG, monogalactosyldiacylglycerol; PA, phosphatidic acid; PC, phosphatidylcholine; PE, phosphatidylethanolamine; PG, phosphatidylglycerol; PI, phosphatidylinositol; PIP and PIP$_2$, phosphatidylinositol mono- and bisphosphate; PS, phosphatidylserine; SE, steryl ester; SG, steryl glycoside; TAG, triacylglycerol.

Prominent alterations of the sterol amounts were observed in PMs from WT plants under CA conditions except for the FS, which remained steady. Namely, increased proportion of SE and decreased proportions of SG and ASG were detected in the WT PM under CA conditions in comparison to NA conditions (Fig. 11). No significant changes were detected in the species profile of FSs in both PM and TE preparations from WT (Fig. 6); only cholesterol was observed to accumulate in higher proportion in TE under CA conditions. Noteworthy, regardless of the sterol core structures of the SE and ASG species, a slight increase of 18:3-containing species and a mild decrease of 18:2-containing species were observed under CA conditions. No significant alterations in the molecular species profile of SGs were detected under CA conditions, although their overall abundance decreased significantly (Fig. 11).

The relative amounts of glycerolipids change in a class-specific manner under CA conditions. Phosphoinositides (PI, PIP and PIP$_2$), PG, PA, DGDG and TAG are elevated; whereas, PC, PE, PS, MGDG and DAG are reduced proportionally in the PM fractions isolated from WT plants (Fig. 11). Although the monounsaturated lipids still denote the most abundant lipid species within all glycerophospholipid classes, they are partially replaced by diunsaturated lipids (Fig. 7). This remodeling happened in both WT PM and TE, especially within the PC, PE, PS and PA classes. In contrast to the other glycerophospholipids, the molecular species profile of PG displays unique alterations, as 16:0/18:3 moieties are included but 16:0/16:1 and 16:0/18:1 species downregulated. Moreover, PG (16:1/18:3) decreases under CA condition in the TE but not in the PM of WT. This particular PG moiety constitutes the predominant PG variant within plastidial membranes, which implies that the observed changes in the TE may result from remodeling of lipid species within the chloroplasts. Noteworthy, although CA greatly induces the accumulation of phosphoinositides in the PM (Fig. 11), their relative proportion remains stable, with only PI (18:3/18:3) displaying a slight increase under CA (Fig. 7). Although the overall amount of DAG and MGDG was reduced, no drastic alterations were observed in its species profile (Fig. 8-9). Significant increases of diunsaturated species including DGDG (16:3/18:3) and DGDG (18:3/18:3) contribute to the elevated DGDG levels observed in PM of WT plants grown under CA conditions. Increased unsaturation degrees can also be observed in TAGs. The species which contain more than six double bonds accumulated while others reduced under CA conditions in both WT PM and TE (Fig. 9). To sum up, the unsaturation degrees of all lipid categories (sphingolipids, sterols and glycerolipids) of WT PMs increased under cold acclimation.

Malfunction of sphingolipid α-hydroxylases does not impair the lipid modulation occurred under cold acclimation in plasma membranes at the species level

The effects of diminished hydroxylated sphingolipids on the PM lipid composition under CA conditions was evaluated based on the detailed lipid profiles of the *fah1 fah2* PM. Here, CA conditions led to the enrichment of LCB and GlcCer, which was not observed in the PM of WT (Fig.

11). On the other hand, other sphingolipids such as LCB-P, Cer, series A and B GIPCs were similarly reduced as in WT, although to a different extent. The overall LCB and LCB-P lipid classes show a strong reduction in PMs from CA *fah1 fah2*, but only mild decreases of LCB (18:0;2) and LCB-P (18:0;3) were detected (Fig. 3). Concerning complex sphingolipids, the characteristic lipid profiles of *fah1 fah2* plants can still be observed under CA conditions; namely, the non-hydroxylated lipids being the predominant species. It should be noted that although the molecular structures of the complex sphingolipids in *fah1 fah2* plants are highly affected by the loss of α-hydroxylases, their responses towards CA conditions remain the same as observed in WT plants: lipid species with unsaturated acyl chains such as cCer (18:1;3/24:1) accumulate while those with saturated acyl chains such as cGlcCer (18:1;3/22:0) are reduced in both PM and TE from CA *fah1 fah2* mutants (Fig. 4). Similar to WT extracts, higher levels of hCers with LCB (18:2;2) backbones including hCer (18:2;2/24:0) and hCer (18:2;2/26:0) were detected. Series A GIPC displayed smaller decreases while series B GIPC displayed greater reduction in *fah1 fah2* PMs in comparison to the respective PMs from WT plants grown under CA conditions (Fig. 11). However, the composition of the detected GIPC species remained steady, with only a slight decrease of cH-H-GIPC (18:1;2/20:0) (Fig. 5). As seen for the other complex sphingolipids, the proportion of GIPCs with LCB (18:2;2) backbones increase in *fah1 fah2* PM under CA conditions.

In contrast to WT, detailed lipid profiling displayed that the overall abundance of FS and SE increased in PMs from *fah1 fah2* propagated under CA conditions; however, SG and ASG remain unchanged (Fig. 11). In general, the occurrence and the predominant species of the detected sterol classes remain the same as under NA conditions. Also, no significant changes were observed regarding the FS species (Fig. 6) and only a slight increase of campesteryl glycosides and decrease of sitosteryl glycosides. Moreover, 18:3-containing SEs and ASGs accumulate while the 18:2-containing species are reduced during CA conditions, which is similar to WT. Noteworthy, fewer alterations in the lipid composition of sphingolipids and sterols were observed in *fah1 fah2* PM under CA condition (Fig. 11), which correlates well with the observations that *fah1 fah2* plants contain a less organized PM (Lenarčič *et al.* 2017).

More prominent alterations were observed in the glycerolipid composition under CA conditions when compared to the NA conditions. In the mutant, CA conditions lead to the loss of PC and PS, whereas PG, phosphoinositides, DGDG, DAG and TAG amounts were elevated (Fig. 11). However, the proportions of the diunsaturated phospholipids still accumulate, and the monounsaturated phospholipids remain reduced but nevertheless represent the major species within all detected glycerophospholipid classes (Fig. 7). Noteworthy, phosphoinositides accumulate greatly in the PMs of both WT and mutant plants under CA conditions (Fig. 11). In addition, slight increases of

PI (18:3/18:3) and PIP$_2$ (18:3/18:3) were detected especially in the *fah1 fah2* PM under CA conditions (Fig. 7). Significant increases of DGDG (16:3/18:3), DGDG (18:3/18:3) as well as polyunsaturated TAGs were observed for both (Fig. 8-9). Overall, although the molecular structures of the complex sphingolipids in *fah1 fah2* plants are highly affected by the loss of α-hydroxylases, the responses of sphingolipids as well as sterols and glycerolipids towards CA conditions correlate well as observed in the WT plants: CA conditions induced the accumulation of lipid species with higher unsaturation degrees.

Individual lipid species are asymmetrically distributed across the plasma membrane

To address the transversal lipid distribution in *Arabidopsis* PM, we generated two populations of WT PMs exhibiting the "apoplastic-side-out" and "cytoplasmic-side-out" orientations with the assistance of a vesicle-inverting detergent, Brij58. The lipid composition of their exterior leaflets was subsequently monitored after chemical and enzymatical treatments. The distribution of the glycolipids including GlcCers, SGs and ASGs was assessed chemically with periodate; phospholipids including PCs, PEs, PSs and PIs were hydrolyzed enzymatically by phospholipase A$_2$. We also attempted to reveal the distribution of GIPCs with periodate; however, the experimental parameters need to be optimized further to enhance the recovery rate.

The distribution of GlcCers in the PM varies according to the molecular species. The most abundant GlcCer species, hGlcCer (18:1;3/24:1), is allocated mainly to the apoplastic leaflet (60 %) (Fig. 12a). In average, 67 % of overall GlcCer species (both cGlcCers and hGlcCers) are found within the apoplastic leaflet (Tab. S4). It should be noted that although GlcCer is proportionally more abundant than other sphingolipid classes, the distribution of other sphingolipids such as Cers and GIPCs may vary and have an impact on the membrane organization as well.

The orientations of the glycosylated sterols, SGs and ASGs, were assessed by periodate treatment. The most abundant SG species are sitosteryl, campesteryl and stigmasteryl glycosides, from which 100 %, 70 % and 18 %, respectively, are allocated to the cytoplasmic side of the *Arabidopsis* PM (Fig. 12b). The distribution of ASGs varies greatly, as the most abundant species, 16:0-sitosteryl glycoside, is located primarily within the apoplastic leaflet (76 %) while sitosteryl glycosides containing either 18:2 or 18:3 fatty acyl moieties are located exclusively within the cytoplasmic leaflet. In average, 74 % and 61 % of the detected SG and ASG species, respectively, are located within the cytoplasmic side of the PM vesicle.

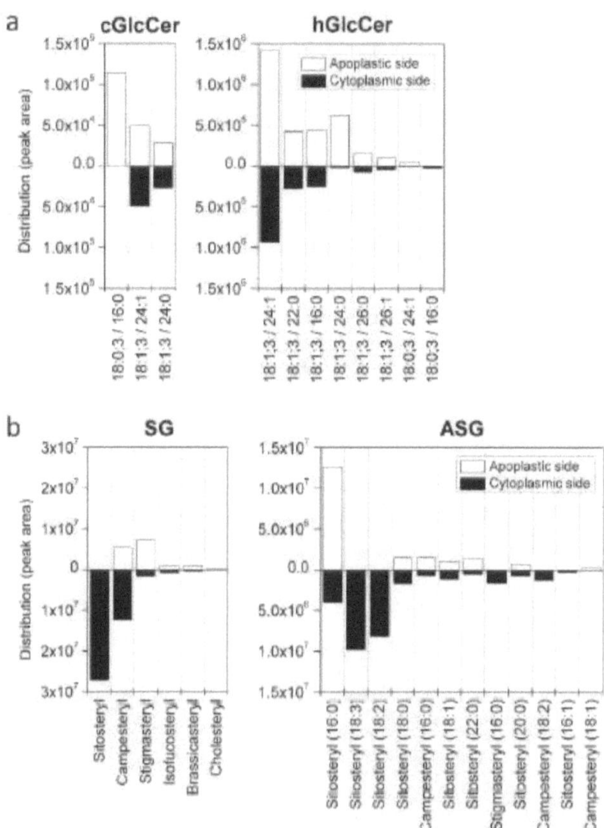

Figure 12. Transversal distribution of glycolipids within the plasma membrane.
Distribution of (a) cGlcCer and hGlcCer, (b) SG and ASG in apoplastic-side-out and chemically inverted cytoplasmic-side-out PM vesicles from wild-type *Arabidopsis* leaves were analyzed by LC-MS/MS after periodate treatment. Data represent the distribution (in peak area) of individual lipid species in the according lipid class from one experiment. ASG, acylated steryl glycoside; cGlcCer and hGlcCer, non- and hydroxylated glucosylceramide; SG, steryl glycoside.

Analysis of the phospholipase A_2-treated PM vesicles revealed that PE (16:0/18:2) and PE (16:0/18:3), the most abundant phospholipid species in *Arabidopsis* PM, are distributed almost symmetrically to both leaflets with 54 % and 63 % locating within the apoplastic side, respectively (Fig. 13). Other PE species display a roughly symmetrical distribution as well, despite 74 % of PE (18:3/20:0) locating within the apoplastic side of the PM vesicles. A higher proportion of PC (16:0/18:2; 78 %) is present within the apoplastic side while the other PC species are allocated almost equally to both leaflets. Most of the identified PS species are located primarily within the cytoplasmic side, which corresponds to previous observations (O'Brien *et al.* 1997). The predominant species, PS (16:0/18:3; 77 %) and PS (16:0/18:2; 65 %), were detected mostly within the cytoplasmic leaflet. In average, 58 % and 68 % of the PE and PE species are distributed within the apoplastic leaflet but 69 % and 53 % of the PS and PI species within the cytoplasmic leaflet, respectively (Tab. S4). Building on the first insight into the PM asymmetry of *Arabidopsis* leaves, additional biological repetitions and further parameter optimizations for the GIPC analysis will provide advanced knowledge concerning the lipid species-specific distribution, the resulting influence on the membrane organization and the lipid-involving signaling pathways within the plant PM.

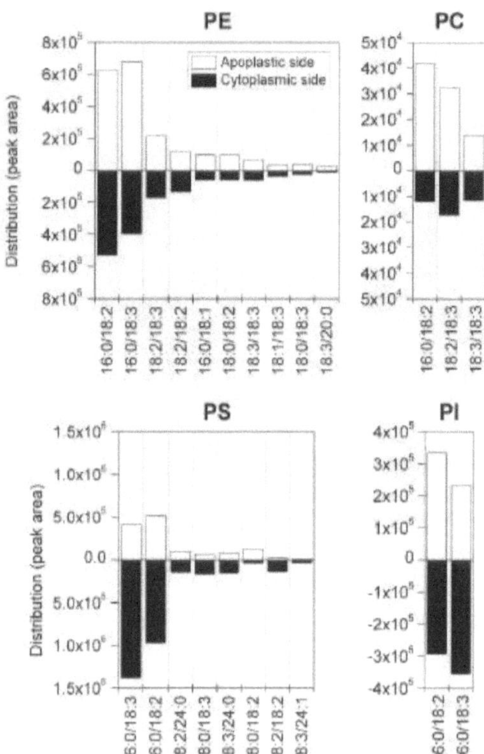

Figure 13. Transversal distribution of glycerophospholipids within the plasma membrane.
Distribution of PE, PC, PS and PI in apoplastic-side-out and chemically inverted cytoplasmic-side-out PM vesicles from wild-type *Arabidopsis* leaves were analyzed by LC-MS/MS after phospholipase A$_2$ treatment. Data represent the distribution (in peak area) of individual lipid species in the according lipid class from one experiment. PC, phosphatidylcholine; PE, phosphatidylethanolamine; PI, phosphatidylinositol; PS, phosphatidylserine.

Discussion

In this study, we present in-depth lipidomics datasets of the PMs from *Arabidopsis* WT and *fah1 fah2* grown under NA and CA conditions with an emphasis on the lipid classes and lipid species of sphingolipids, sterols and glycerolipids. The alterations between the lipid profiles revealed the impacts on the remodeling of the PM lipids caused by the loss of sphingolipid fatty acid α-hydroxylase and by CA. Furthermore, we identified the transversal distribution of the most abundant glycolipids and phospholipids within the WT PMs at the molecular species level. Previous studies have demonstrated that α-hydroxylated sphingolipids are essential in establishing the lipid rafts on the PM, and that the *fah1 fah2* PM is less orderly packed in comparison to the WT PM (Lenarčič *et al.* 2017). However, information on the detailed lipid molecular species of the *fah1 fah2* PM concerning acyl chain length and unsaturation degree of the lipids, which are critical parameters to determine the membranes physical properties, was lacking. In addition, the transversal distribution of individual lipid species across the PM, which influences the surface potential and signal transduction, has not been profiled in the plant system.

Purified plasma membrane was obtained from *Arabidopsis* leaves

In the present study, PMs were purified by two-phase partitioning system, which separates the PM vesicles from other intracellular membranes (Larsson *et al.* 1994). Much effort was made to evaluate the purity of the resulting PM fraction: (i) LFQ shot-gun proteomic analysis in combination with subcellular localization assessed by SUBAcon database, and (ii) in-depth lipidomics analyses, wherein the presence of plastidial lipid molecular species were inspected. The SUBAcon database defines consensus subcellular localizations of Arabidopsis proteins based on its unified data collection of multiple sources of experimental data (microscopy, MS, protein-protein interaction and co-expression) and 22 bioinformatic prediction algorithms. Based on this analysis, purity of the PM fractions can be estimated to be 89 % (Fig. 1c). PM-specific proteins were highly enriched in our PM preparations. Traces of membrane proteins from endomembranes (ER, Golgi apparatus and vacuole), plastids and peroxisomes were present; nevertheless, all these proteins were strongly reduced (negative LFQ values) in the PMs, when compared to other membrane fractions (Tab. S1). Quantitative analysis of the glycerolipids in the PMs revealed that the most enriched lipid class is PE, which corresponds to previous studies (Uemura *et al.* 1995). Minor amounts of MGDG, DGDG and PG were also present in our PMs. Although they are the main components of plastidial lipids, they have also been identified in the PMs as demonstrated previously (Tjellström *et al.* 2010). Distinct species profiles of these lipid classes were identified in our PMs: MGDG, DGDG and PG species with 16:0 fatty acyl moieties were enriched in the PMs, whereas the plastidial-specific species including MGDG (16:3/18:3), DGDG (18:3/18:3) and PG (16:1/18:3) were strongly reduced (Fig. 7-8). This suggests that the contaminants of intact plastids or bulk plastidial membranes are insignificant. The presence of endomembrane-associated proteins was expected as lipids intended for the PMs are being transported through the vesicular transportation pathway, wherein ER and Golgi apparatus are highly involved. Nevertheless, they account for only 6 % in the identified proteins in our PMs. Therefore, we conclude that the overall PM preparations contain only trace subcellular membrane pieces, which may originate from the homogenization process, membrane contact sites between the PMs and other organelles (i.e. plastids) and PM-targeted vesicles derived from the vesicular transportation pathway (i.e. endomembranes). It should be noted that PI-specific phospholipase C (PI-PLC), phospholipase D α1 (PLDα1) and phospholipase D δ (PLDδ), although not enriched, were also identified in our PM proteome. They are involved in the generation of PA and DAG molecules specifically from PI by PI-PLC, preferentially from PC and PE by PLDs (Qin *et al.* 2002, Takáč *et al.* 2019) and may have a slight influence on the composition of the PM lipidome. This influence cannot be omitted from the membrane preparation procedure, since several long ultracentrifugation steps are included to ensure the purity of the isolated PM.

The landscape of the plasma membrane is enriched with raft-forming lipids

It has been demonstrated that the proportion of sphingolipids, sterols and phospholipids in the PMs varies greatly according to the plant species and tissues. For instance, in the PM isolated from leaf tissues, the content of sphingolipids range from 6 % in wild potato to 31 % in winter oat; sterols from 22 % in spinach to 47 % in winter rye; and phospholipids from 29 % in winter oat to 64 % in spinach (Palta *et al.* 1993, Rochester *et al.* 1987, Takahashi *et al.* 2016, Uemura *et al.* 1995, Uemura and Steponkus 1994). A relatively low content of sphingolipids (7 %), but high contents of phospholipids (47 %) and sterols (46 %) were identified in the PM isolated from *Arabidopsis* leaves (Uemura *et al.* 1995). In addition, the sphingolipids containing 24:0 as acyl moiety and the polyunsaturated phospholipids are characteristic for the PM of *Arabidopsis* suspension cells (Grison *et al.* 2015). Here, although the comparison between the absolute lipid abundance in our PMs remains to be determined since the quantitative analyses of sphingolipids and sterols are still in progress, we demonstrated via LC-MS-based lipidomics that a greater lipid variety in both lipid class and molecular species levels is present in the PM as compared to the TE. Significantly larger proportions of the raft-forming lipids including hCers, hGlcCers and series A hGIPCs as well as free campesterol and stigmasterol were detected in the WT PM (Fig. 4-6). However, the molecular species profiles of the other sterol derivatives including SEs, SGs and ASGs are similar between the PM and TE of the WT plants, suggesting that they occur in a certain ratio throughout distinct subcellular membranes (Fig. 6). The predominant phospholipids in the WT PM are the monounsaturated species; the diunsaturated species occur in a smaller amount; and the disaturated species account for the least proportion (Fig. 7). This high unsaturation degree of the PM-localized phospholipids has been observed previously as well (Grison *et al.* 2015). Noteworthy, significantly higher proportions of VLCFA-containing species like PS (18:2/26:0) and PS (18:3/26:0) are located to the WT PM in comparison to the WT TE. We speculate that these lipid species may align closely to the lipid rafts, which are thicker membrane patches enriching in the VLCFA-containing sphingolipids, thereby contribute to their formation.

The lower level of ordered plasma membrane domains caused by the decline of α-hydroxy groups in fatty acids of sphingolipids may be rescued by multi-glycosylated sphingolipids in *fah1 fah2* plasma membrane

As observed in the previous researches, strong reductions of complex sphingolipids are characteristic in the *fah1 fah2* TE (König *et al.* 2012, Lenarčič *et al.* 2017). In this study, reduced levels of complex sphingolipids including Cer, GlcCer and series A GIPC were detected in the *fah1 fah2* PM (Fig. 10). The predominant species in the *fah1 fah2* PM contain fatty acyl chains with either 16, 22 or 24 carbons in length, which is the same as observed in the WT PM, albeit without α-hydroxylation due to the loss of the fatty acid α-hydroxylases (Fig. 4 and 5). The α-hydroxyl groups on complex sphingolipids have been demonstrated to be involved in bridging sphingolipids and sterols to establish lipid rafts (Löfgren and Pascher 1977, Pascher and Sundell 1977). In addition, the malfunction of the α-hydroxylases in *fah1 fah2* mutants leads to reduced levels of ordered plasma membrane and strong reduction of GlcCer and GIPC levels in leaf total extracts of *Arabidopsis* (König *et al.* 2012, Lenarčič *et al.* 2017). Here, we confirmed that all raft-forming lipid classes, including GlcCer, series A GIPC, SG and ASG, display a strong reduction in both PM and TE of *fah1 fah2*, with even more profound results in the *fah1 fah2* PM (Fig. 10). Interestingly, considerably higher proportions of series B GIPC (both non- and hydroxylated species) are present in the *fah1 fah2* PM. It has been demonstrated that in mammalian cells and marine sponges, glycolipids are able to interact with other multi-glycosylated molecules such as glycoproteins and surface glycan to form glycoconjugates on the PM (Bucior and Burger 2004, Handa and Hakomori 2017, Popescu *et al.* 2003). These glycoconjugates stabilize the PM organization and additionally mediate the binding of pathogenic bacteria to the mammalian host cells (Day *et al.* 2015). Furthermore, the series B GIPCs in the *fah1 fah2* PM carry shorter fatty acyl moieties (20 carbon) (Fig. 5), which are suitable for the less-ordered and raft-depleted *fah1 fah2* PM. These results lead us to hypothesize that the increase of multi-glycosylated molecules, but not the elongation of fatty acyl chain length, may rescue the decline membrane ordered domains caused by the absence of α-hydroxy groups in fatty acids of sphingolipids. It has been demonstrated that SEs are a highly assessable reservoir for maintaining the homeostasis of the FSs on the plant PM (Bouvier-Nave *et al.* 2010, Lara *et al.* 2018). As a result, a proportionally higher SE level was detected in the *fah1 fah2* PM while the FS amount remains constant (Fig. 10), suggesting that the sterol biosynthesis could be redirected to generate SEs instead of SGs and ASGs to maintain membrane organization. A strong accumulation of LCB-Ps, especially LCB-P (18:1;3), was also detected in both PM and TE of *fah1 fah2* (Fig. 3). It is appealing to propose that LCB-Ps serves as a sink for an overproduction of sphingolipid precursors and can constitute an alternative product for the excess LCBs, which cannot be converted further to complex sphingolipids. However, a high LCB-P to LCB ratio can increase the sensitivity towards abscisic acid and modulate the LCB-induced program cell death (PCD) in *Arabidopsis* (Alden *et al.* 2011, Berkey *et al.* 2012, Guo *et al.* 2011, Saucedo-García *et al.* 2011, Shi *et al.* 2007, Tsegaye *et al.* 2007, Worrall *et al.* 2008). Activated PCD responses as well as the elevated SA levels have already been detected in *fah1 fah2* plants leading to inhibited growth (König *et al.* 2012), and further investigations are required to assess the correlation between the sphingolipid biosynthesis pathway, the LCB-induced PCD and the SA responses.

The loss of sphingolipid α-hydroxylases and cold stress may trigger similar responses in *Arabidopsis* leaves

Typical cold-induced responses including the depletion of raft-forming sphingolipids and sterols as well as the accumulation of PA (Minami *et al.* 2008, Ruelland *et al.* 2002) have been observed in both WT PMs under CA conditions and *fah1 fah2* PMs under NA conditions. This suggests that the loss of α-hydroxyl groups in fatty acids of sphingolipids and CA may trigger similar responses in *Arabidopsis* leaves. Cold-induced responses include numerous other physiological and biochemical modifications such as Ca^{2+} signaling, post-transcriptional and post-translational modifications, chloroplast status and phytohormone alterations (Miura and Furumoto 2013, Pál *et al.* 2013). Noteworthy, it has been demonstrated that higher levels of endogenous SA and its derivatives, especially the glycosylated form, accumulate during CA condition and enhance the cold tolerance in several plant species (Kosová *et al.* 2012, Scott *et al.* 2004). Indeed, constitutively elevated levels of SA and its derivatives were also detected in *fah1 fah2* plants under NA conditions (König *et al.* 2012). Furthermore, an enhanced resistance towards biotrophic pathogens was observed in *fah1 fah2* plants (König *et al.* 2012), which corresponds not only to the high accumulation of SA, but also to the strong reduction of their complex sphingolipids known to be involved in the pathogenic recognition (Lenarčič *et al.* 2017). Notably, in contrast to the changes in the WT PM under CA conditions, a notable accumulation of DAG was detected in the *fah1 fah2* PM under CA conditions (Fig. 11). Since DAG is a by-product of inositol phosphoceramide synthase during sphingolipid biosynthesis (Lee and Assmann 1991), the accumulation of DAG may result from the impaired sphingolipid biosynthesis metabolism in *fah1 fah2* plants.

Here, CA leads to an increase in phospholipids commonly observed in many herbaceous and woody plant species (Palta *et al.* 1993, Rochester *et al.* 1987, Takahashi *et al.* 2016, Uemura *et al.* 1995, Uemura and Steponkus 1994). Accumulation of PG and phosphoinositides and concomitant reduction of PC and PS were observed in both WT PM and *fah1 fah2* PM under CA conditions (Fig. 11). Under cold and osmotic stresses, phosphoinositides play roles in modulating the electrostatic signature and reorganizing the cytoskeleton, thus affecting the membrane structure (DeWald *et al.* 2001, Pical *et al.* 1999, Smoleńska-Sym and Kacperska 1994, Williams *et al.* 2005). In addition, phosphoinositides work cooperatively with PS, as well as other anionic phospholipids such as PA, to establish the membrane electrostatics of plant PM (Platre *et al.* 2018). Therefore, the drastic increase of these anionic phospholipids may contribute to the reduction of PS under CA conditions in order to balance the surface charge. Under CA conditions, higher levels of diunsaturated species and lower levels of monounsaturated species occur in all the detected phospholipid classes to increase the membrane fluidity in both WT PM and *fah1 fah2* PM (Fig. 7). It should be noted that increasing diunsaturated PC is hypothesized to play a role in the development of freezing tolerance as well (Inatsugi *et al.* 2002, Uemura *et al.* 1995, Uemura and Steponkus 1994). In addition, glyceroglycolipids, neutral glycerolipids, complex sphingolipids and sterol derivatives that contain fatty acyl moieties with higher unsaturation degrees accumulated under CA conditions in both WT PM and *fah1 fah2* PM (Fig. 4-6, 8), which may collectively contribute to the modulation of PM fluidity. Previous researches have indicated that DGDG increases under low temperature (Gu *et al.* 2017, Welti *et al.* 2002), and can be relocated to extraplastidial membranes including the PMs under phosphate-limiting conditions (Andersson *et al.* 2005, Andersson *et al.* 2003, Jouhet *et al.* 2004, Tjellström *et al.* 2010). The accumulation of specific DGDG species in the PMs may contribute to the cold-induced responses in correlation to the remodeling of phospholipids.

GlcCer, PE and PC are distributed preferentially to the apoplastic leaflet while SG, ASG, PS and PI to the cytoplasmic leaflet

It has been presumed that the plant PM establishes a similar transversal lipid asymmetry as observed in animal cells and in yeasts (Devaux *et al.* 2006, Hill *et al.* 1999, Krylov *et al.* 2001). In addition, a recent study indicated that more unsaturated phospholipids are distributed to the cytoplasmic leaflet in comparison to the exoplasmic leaflet of the mammalian PM (Lorent *et al.* 2020). In contrast to yeasts and animal systems, the plant PM contains a wide variety of sphingolipids, FSs and sterol derivatives. However, only a few studies have focused on the PM lipid asymmetry in plants. It has been demonstrated that 70 % of GlcCer, 70 % of total sterols and 35 % of phospholipids are distributed to the apoplastic leaflet in the PM of oat roots (Tjellström *et al.* 2010); 98 % of GlcCer are distributed to the apoplastic leaflet in the PM of squash leaves (Lynch and Phinney 1995); and phospholipids are distributed symmetrically between the PM leaflets in hypocotyl cells of mung beans (Takeda and Kasamo 2001). Here, we demonstrate that individual lipid molecules are not exclusively allocated to one leaflet, instead, they reach a homeostasis between the two leaflets of the PM. About 60 % of the predominant GlcCer species, hGlcCer (18:1;3/24:1), is distributed to the apoplastic leaflet (Fig. 12a) and the most abundant SG species, sitosteryl glycoside, is located exclusively within the cytoplasmic leaflet of the WT PM (Fig. 12b). Although SG has been indicated to act as the primer for the exterior-orientated cellulose synthesis (Peng *et al.* 2002), individual SG species may exert distinct functions on the different leaflets of the PM. The distribution of individual SG and ASG species varies greatly, suggesting that their different roles in membrane organization may be determined by both the sterol cores and the acyl chains (Lefebvre *et al.* 2007). It has been proposed that SG and ASG can modulate the activity of ATPases on the PM, and thus regulate the enzyme-associated lipid rafts (Bhat and Panstruga 2005, Lefebvre *et al.* 2007, Mongrand *et al.* 2004). However, the influence of individual lipid species on the organization of the plant PM and their association with lipid rafts is still elusive.

To access the transversal distribution of phospholipids, phospholipase A_2 that hydrolyzes the β-ester bond of PE, PC, PI and PS was applied to demonstrate their distribution within the plant PM. PE is the most abundant phospholipid class in the WT PM (Fig. 2). In average, PE and PC are distributed preferentially to the apoplastic leaflet (58 % and 68 %, respectively); whereas, anionic phospholipids including PS and PI are distributed to the cytoplasmic leaflet (69 % and 53 %, respectively) (Fig. 13). Noteworthy, although PS is distributed almost exclusively to the cytoplasmic leaflet of the mammalian PM, it can still be detected on the apoplastic leaflet of the *Arabidopsis* PM. This observation is supported by a previous study, which also detects PS on the apoplastic side of tobacco protoplasts via PS-binding protein, annexin V (O'Brien *et al.* 1997). In average, phospholipids display a roughly symmetric distribution on the PM bilayer as observed previously (Takeda and Kasamo 2001), resulting from the preferential distribution of PC and PE within the apoplastic leaflet and PI and PS within the cytoplasmic leaflet.

In summary, we expanded the knowledge regarding the membrane organization of the plant PM by determining the absolute abundance of glycerolipid classes and providing in-depth molecular species profiles of sphingolipids, sterols and glycerolipids in leaves of the model plant *Arabidopsis*. The quantitative analysis demonstrated that PE is the most abundant glycerolipid class in plant PM. In addition, molecular species of sphingolipids with α-hydroxylated fatty acid moieties, sterols with sitosterols and campesterols as the core structures and glycerophospholipids with a single unsaturated fatty acyl moiety per lipid species are enriched in WT PM. The α-hydroxylase mutations on the *fah1 fah2* PMs lead to strong reductions of hydroxylated complex sphingolipids and sterols, but an accumulation of series B GIPCs. Noteworthy, we demonstrated that the impact of the α-hydroxylase mutations on the *fah1 fah2* PM resembles the cold-induced responses in the WT PM, including the reduction of complex sphingolipids and sterols and the accumulation of PAs. Nevertheless, mutant plants are still capable of modulating the membrane biophysical properties as the WT plants under CA condition by accumulating lipid species with higher unsaturation degrees to increase the membrane fluidity under cold stress. Although hydroxylated complex sphingolipids are diminished in the *fah1 fah2* mutant plants, the non-hydroxylated complex sphingolipids display a similar tendency with species with higher unsaturation degrees accumulating under CA conditions. Furthermore, our study provides first preliminary data on the species-specific transversal distribution of the most abundant GlcCer, SG, ASG and phospholipids in the *Arabidopsis* PM. Molecular species of GlcCers, PEs and PCs are distributed preferentially to the apoplastic leaflet (67 %, 58 % and 68 %, respectively) while SGs, ASGs, PSs and PIs to the cytoplasmic leaflet (74 %, 61 %, 69 % and 53 % respectively). Noteworthy, we observed that individual lipid species, even from the same lipid class, are distributed differently on the two leaflets of the *Arabidopsis* PM, suggesting that individual lipid species may exert specific functions involved in the lipid signaling and membrane organization.

Materials and methods

Plant material and growth condition

Wild type (WT) and *fah1 fah2* mutant line were all in the *Arabidopsis thaliana* (L.) Heynh Columbia-0 background. Homozygous plants of *fah1 fah2* line were described before (König *et al.* 2012). After sowing the seeds in pots with 3-d cold stratification at 4 °C, seedlings were grown under NA conditions with 16-h light at 23 °C for one week. Young seedlings were transferred to large trays and grown with equal spacing for another two weeks. CA was achieved by reducing the temperature to 6 °C for additional 10 days with 12 h light.

Plasma membrane isolation

For every membrane preparation, 100 g of *Arabidopsis* rosettes were collected and homogenized in a kitchen blender in 200 ml of ice-cold homogenization buffer (50 mM MOPS, pH 7.0; 2 mM EGTA; 2 mM EDTA; 0.1 mM PMSF). The homogenate was filtered through two layers of miracloth with supporting gauze and centrifuged at 10,000 g at 4 °C for 15 min. The supernatant, which includes the microsomal membranes, was collected and centrifuged at 50,000 g at 4 °C for 60 min to generate microsomal pellets. The microsomal fraction (MC) was obtained after resuspending the microsomal pellets with 5 mM potassium phosphate buffer, pH 7.6 and homogenizing them with a 27 G needle. For PM enrichment, aliquots of 6 mL of 4-8 mg ml^{-1} MC were transferred to 18 g two-phase solutions (5.8 % Dextran 500, w/w; 5.8 % PEG 3350, w/w; 5 mM potassium phosphate buffer, pH 7.6; 2 mM KCl). The PM was collected after two-phase partitioning as described previously (Larsson *et al.* 1994). In brief, thoroughly mixed two-phase solutions were centrifuged at 648 g at 4 °C for 10 min: upper and lower phases harbor PM and other IM, respectively. To enhance the purity of the final PM fractions, the phase partitioning was repeated twice. IM- and PM-containing phase solutions were transferred to clean tubes and diluted with five-time volume of water or dilution buffer (10 mM HEPES/KOH, pH 7.5; 0.25 M sucrose; 10 mM KCl). IM and PM were pelleted after centrifugation with 100,000 g at 4 °C for 60 min, and resuspended with 0.8 ml dilution buffer before storing at -20 °C.

Protein determination

Estimation of protein concentration was conducted as previously described (Esen 1978). In brief, filter papers (FN6) harboring 5 µl membrane samples or 2 mg ml^{-1} BSA solution were incubated with staining solution (25 % propan-2-ol, v.v; 10 % acetic acid, v.v; 0.05 % Coomassie blue G250, w/v) for 20 min. Excess dye from the filter papers was eliminated by two subsequent washing steps with alternating room-temperature and boiling water. The protein – dye conjugation was released in 3 ml 0.5 % (w/v) SDS by 50 °C incubation for 20 min. The absorbance at 578 nm of the resulting dye-containing solutions was measured and utilized for calculating the sample concentration in comparison to the BSA standard.

Lipid extraction

Total lipid extract from leaf was obtained as described (Tarazona *et al.* 2015). Briefly, mixture of 150 mg frozen leaf powder and 6 ml extraction buffer (propan-2-ol : hexane : water, 60:26:14, v.v.v) was incubated at 60 °C with shaking for 30 min. After centrifugation at 800 g for 20 min, the clear supernatant was evaporated under a stream of nitrogen gas until dryness. Samples were reconstituted with 0.8 ml of tetrahydrofuran : methanol : water (TMW, 4:4:1, v.v.v) for the subsequent LC-MS/MS measurements. The procedure for extracting the membrane lipids was adjusted by substituting the water fraction of the extraction buffer by equal volume of membrane samples. Aliquots of 750 µL PM, 780 µL IM and 190 µL MC fractions were used in this study. Further process was continued as above.

Lipid derivatization

Phosphate-containing lipids –PA, PIP and PIP$_2$ and LCB-P – were derivatized to enhance their chromatographic separation and mass spectrometric detection. Methylation of PA, PIP and PIP$_2$ was performed as described with slight modifications (Lee *et al.* 2013). Aliquots of 100 µl lipid extracts were first dried under a stream of nitrogen gas and reconstituted with 200 µl methanol. Methylation reaction took place after adding of 3.3 µl trimethylsilyldiazomethane (2 M in hexane, Merck KgaA, Darmstadt, Germany). After 30 min incubation at room temperature, the reaction was terminated by neutralizing the solution with 1 µl of 1.7 M acetic acid. Samples were dried under nitrogen gas and redissolved in 100 µl TMW. Acetylation of LCB-P was performed as described with slight modifications (Berdyshev *et al.* 2005). Aliquots of 100 µl lipid extracts were dried and reconstituted with 100 µl pyridine and 50 µl acetic anhydride. After 30 min incubation at 50 °C, samples were dried under a stream of nitrogen gas at 50 °C. To redissolve lipids, 100 µl TMW was used as the final solvent to proceed with the lipid analysis.

Semi-quantitative lipidomics analysis with LC-MS

Analysis conditions and system setup were similar to our previous work (Tarazona *et al.* 2015). Samples were separated by an ACQUITY UPLC system (Waters Crop., Milford, MA, USA) with a HSS T3 column (100 mm x 1 mm, 1.8 µl; Waters Crop.), ionized by a chip-based nano-electrospray using TriVersa Nanomate (Advion BioScience, Ithaca, NY, USA) and analyzed by a 6500 QTRAP tandem mass spectrometer (AB Sciex, Framingham, MA, USA). Aliquots of 2 µl were injected and separated with a flow rate of 0.1 ml min^{-1}. The solvent system was composed of methanol : 20 mM ammonium acetate (3:7, v.v) with 0.1 % acetic acid, v.v (solvent A) and tetrahydrofuran : methanol : 20 mM ammonium acetate (6:3:1,v.v.v) with 0.1 % acetic acid, v.v (solvent B). According to the lipid classes, different linear gradients were applied: start from 40 %, 65 %, 80 % or 90 % B for 2 min; increase to 100 % B in 8 min; hold for 2 min and re-equilibrate to the initial conditions in 4 min. Starting condition of 40 % solvent B were utilized for LCB and LCB-P; 80 % for DAG; 90 % for TAG and SE; 65 % for the remaining lipid classes. Subsequent retention time alignment and peak integration were performed with MultiQuant (AB Sciex). Further data processing and statistics were done in Excel. The lipid class profiles were generated based on the relative proportion of lipid class-specific sums of peak area.

Proteomic sample preparation

Aliquots containing 10 to 50 µg proteins were used to prepare the samples for proteomic analysis. Proteins were delipidated and precipitated in ten-time volume of 96 % ethanol at − 20 °C. The protein pellets were redissolved in 50 µl SDS protein sample buffer (40 mM Tris, pH 6.8; 1 %SDS, w/v; 5 % glycerol, v.v; 0.003 ‰ bromophenol blue, w/v; 50 mM DTT). An aliquot of 10 µg protein was loaded on to an SDS-PAGE gel. The electrophoresis was stopped when the proteins entered the separating gel for 1 cm. The sample-containing gel piece was cut out and sliced into 0.5 cm × 1 cm pieces to proceed with tryptic digestion and derivatization as described (Shevchenko *et al.* 2006). Peptides were desalted via EmporeOctadecyl C18 47-mm extraction disks 2215 (Merck KgaA, Darmstadt, Germany) (Rappsilber *et al.* 2007), and dissolved in 20 µl 0.1 % formic acid, v.v, before LC-MS/MS analysis.

Proteomic analysis with LC-MS/MS

Analysis condition and system setup for the proteomic analysis were as described (Schmitt *et al.* 2017). Samples were injected into a RSLCnano Ultimate 3000 system (Thermo Fisher Scientific, Waltham, MA, USA) liquid chromatographic system with an Acclaim PepMap 100 precolumn (100 µm × 2 cm, C18, 3 µm, 100 Å; Thermo Fisher Scientific) at a flow rate of 20 µl min^{-1} for 3 to 6 min; then separated by an Acclaim PepMap RSLC column (75 µm × 50 cm, C18, 3 µm, 100 Å; Thermo Fisher Scientific) at a flow rate of 300 nl min^{-1}.The solvent system was composed of 0.1 %formic acid, v.v (solvent A) and 80 % acetonitrile with 0.1 % formic acid, v.v (solvent B). A linear gradient from 2 % solvent B to 32 % B was applied for 94 min, and continues to 65 % B for additional 26 min. The samples were ionized online by Nanospray Flex Ion Source (Thermo Fisher Scientific) and analyzed with an OrbitrapVelosPro hybrid mass spectrometer (Thermo Fisher Scientific). Full scans were executed by its Orbitrap-FT analyzer at a resolution of 30,000 with the mass range from 300 to 1850 m/z; and the data-dependent acquisition of top 15 features (dynamic exclusion enabled) was performed with its Velos Pro analyzer.

MS/MS data processing including peptide analysis, protein identification and label-free quantification was performed by MaxQuant 1.6.1.0 (Cox and Mann 2008). Label-free quantification (LFQ) algorithm was implemented in MaxQuant software (Schaab *et al.* 2012) and applied to determine the protein abundance in different membrane fractions. The false discovery rate of peptide – spectrum matches and proteins were set to 0.02, matches between runs and dependent peptides were enabled, default settings were used for other parameters. The output data were visualized with Perseus 1.6.1.1 (Tyanova *et al.* 2016).

Thin layer chromatographic separation of glycerolipids

Lipid extracts of membrane fractions were spotted onto TLC 60 plates (20 × 20 cm, Merck KgaA) in parallel with corresponding standards. The TLC method was performed as described (Wang and

Benning 2011) with minor modifications. In brief, the TLC plate was pretreated with 0.15 M ammonium sulfate for 30 sec and dried for 2 days at room temperature. Prior to loading the samples, the plate was activated by 120 °C for 2.5 h. Lipid classes were separated in a solvent mixture of acetone : toluene : water (91:30:8, v.v.v). All lipid spots on TLC plates were visualized with 0.5 mg ml⁻¹primuline in the mixture of acetone : water (8:2, v.v) (White *et al.* 1998) under 528 nm UV light, scraped off and converted to fatty acid methyl esters (FAME) by transesterification before GC analysis.

Transesterification

Glycerolipids were converted to FAME by acidic methanolysis (Miquel and Browse 1992). The lipid-bound silica powder from corresponding spots on TLC plate were scraped off and added to 1 ml FAME solution (methanol : toluene : sulfuric acid : dimethoxypropane, 33:17:1.4:1, v.v.v.v) with 5 µg triacylglycerol, TAG (15:0/15:0/15:0), as the internal standard. After 1 h incubation at 80 °C, 1.5 ml saturated NaCl solution and 1.2 ml hexane were added, followed by vigorous mixing. The hexane phase containing the resulting FAME s was, dried, and FAMEs reconstituted in 10 µl acetonitrile.

GC-FID analysis

After converting them to FAMEs, lipid-bound fatty acids were analyzed by a 6890N Network GC-FID system with a medium polar cyanopropyl DB-23 column (30 m × 250 µm × 25 nm; Agilent Technologies, Waldbronn, Germany) using helium as the carrier gas at 1 ml min⁻¹. Samples were injected at 220 °C with an Agilent 7683 Series injector in split mode. After 1 min at 150 °C, the oven temperature raised to 200 °C at the rate of 8 °C min⁻¹, increased to 250 °C in 2 min, and held at 250 °C for 6 min.

FS were analyzed by a similar system setup after silylation by *N,O*-bis(trimethylsilyl)trifluoroacetamide (BSTFA; Merck KgaA), but with a nonpolar HP-5 column (30 m × 250 µm × 25 nm; Agilent Technologies) at a flow rate of 1 ml min⁻¹. Samples were injected at 250 °C in split mode. After 1 min at 220 °C, the oven temperature was raised to 325 °C at the rate of 20 °C min⁻¹ and held for 7.5 min.

Fatty acids and sterols were identified according to the corresponding standards or by additional mass spectrometric identification. Peak integration was performed using the GC ChemStation (Agilent Technologies).

Membrane asymmetry determination

The inversion of PM vesicle was performed as described (Johansson *et al.* 1995) with modifications. In short, cytoplasmic-side-out vesicles were generated by incubating the PM fraction with 0.33 %, v.v, Brij58-containing dilution buffer (10 mM HEPES/KOH, pH 7.5; 0.25 M sucrose; 10 mM KCl) on ice for 30 min. After pelleting the cytoplasmic-side-out vesicles at 10,000 g for 1 h, PM vesicles with both orientations (with and without inversion by Brij58) were treated with PLA_2 to enzymatically digest phospholipids or with sodium periodate to chemically derivatize glycolipids. An aliquot of PLA_2 (1.0 U per 200 µg membrane protein) from porcine pancreas was applied to the PM vesicles at 30 °C for 0 min and 30 min to hydrolyze the *sn*-2 esters of phospholipids, especially PC, PE, PS and PI. The enzymatic reaction was terminated by adding a solvent mixture of methanol : chloroform : formic acid (20:10:1, v.v.v) for the subsequent lipid extraction. To address the transversal distribution of glycolipids including GlcCer, SG and ASG, sodium periodate was applied to the PM vesicles to the final concentration of 10 mM. The chemical treatment was terminated after 0 h and 1 h incubation at 30 °C in the dark by adding the solvent mixture for lipid extraction as mentioned. Subsequent lipid extraction was performed as described previously (Wang and Benning 2011) with an additional repeat of the phase partitioning procedure. Lipid extracts from 0-min samples serve as non-treated controls in the lipid analyses.

The membrane asymmetry was calculated based on the knowledge that the apoplastic-side-out and the cytoplasmic-side-out PM vesicle populations are 85 % and 100 % of the assigned orientation in average, respectively (Johansson *et al.* 1995, Larsson *et al.* 1994). That is, 85 % of the reduced signals in the treated apoplastic-side-out vesicles in comparison to the non-treated extracts result from the modified apoplastic-localized lipids. On the other hand, 100 % of the reduced signals in the treated cytoplasmic -side-out vesicles in comparison to the non-treated extracts result from the modified cytoplasmic-localized lipids.

Acknowledgement

We are grateful for the excellent technical assistance from Sabine Freitag, Susanne Mester, Maria Paulat and Pia Meyer, and helpful discussions with Dr. Amelie Kelly. YL has been a doctoral student of the Ph.D. program "Molecular Biology" – International Max Planck Research School and the Göttingen Graduate School for Neurosciences, Biophysics, and Molecular Biosciences (GGNB) (DFG grant GSC 226) at the Georg August University Göttingen. IF and OV acknowledge funding through the German Research Foundation (DFG: INST 186/822-1, INST 186/1167-1, INST 186/1230-1 and ZUK 45/2010).

Table S1. Overall identified proteins of all membrane fractions via proteomic analysis with label-free quantification.

Table S2. Overall identified lipid species of plasma membrane fractions and total extracts of *Arabidopsis* WT and *fah1 fah2* leaves.

Table S3. Fold changes of lipid classes of (A) plasma membrane fractions (PM) and total extracts (TE) between *Arabidopsis* wild type (WT) and *fah1 fah2* (*fxf*) under non-acclimated condition (B) PM isolated from WT and *fxf* plants between cold-acclimated (CA) and non-acclimated (NA) conditions.

Table S4. Transversal lipid species distribution of glycolipids and phospholipids.

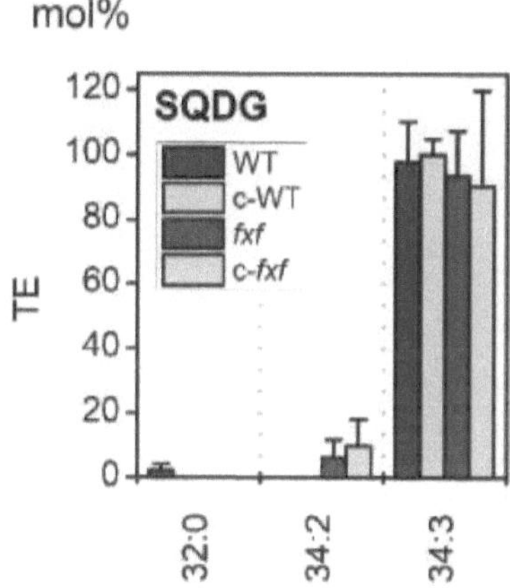

Figure S1. Sulfoquinovosyldiacylglycerol composition of Arabidopsis WT and *fah1 fah2* plants grown under non- and cold-acclimated conditions.
Sulfoquinovosyldiacylglycerols of total extracts (TEs) from wild-type and *fah1 fah2 Arabidopsis* grown under normal (WT and *fxf*) and cold-acclimated (CA) conditions (c-WT and c-*fxf*) were analyzed by LC-MS/MS. Data represent mean values of mol % of an individual lipid species in the according lipid class from three independent experiments ± SD. Sulfoquinovosyldiacylglycerol, SQDG.

Chapter 5.

Defining the lipidome of Arabidopsis leaf mitochondria: Specific lipid complement and lipid biosynthesis capacity

The article was submitted in June 2020. The supplemental materials are attached at the end of the chapter.

Author contribution

Yi-Tse Liu performed the isolation of mitochondria from *Arabidopsis* leaves, extraction of membrane lipids, analysis and subsequent data processing of the TLC-GC and LC-MS/MS results. Moreover, she cross compared the datasets resulting from lipidomics and proteomics analyses as well as online databases to construct the lipid biosynthesis pathways in plant mitochondria. She analyzed, processed, displayed and discussed the data resulting from those experiments, and wrote the first version of the manuscript.

Defining the lipidome of Arabidopsis leaf mitochondria: Specific lipid complement and lipid biosynthesis capacity

Yi-Tse Liu[1], Jennifer Senkler[2], Cornelia Herrfurth[1,3], Hans-Peter Braun[2], Ivo Feussner[1,3,4,*]

[1]University of Goettingen, Albrecht-von-Haller-Institute for Plant Sciences, Department of Plant Biochemistry, 37077 Goettingen, Germany.

[2]Leibniz Universität Hannover, Institute of Plant Genetics, Department of Plant Proteomics, 30419 Hannover, Germany.

[3]University of Goettingen, Goettingen Center for Molecular Biosciences (GZMB), Service Unit for Metabolomics and Lipidomics, 37077 Goettingen, Germany.

[4]University of Goettingen, Goettingen Center for Molecular Biosciences (GZMB), Department of Plant Biochemistry, 37077 Goettingen, Germany

*Correspondence: Ivo Feussner, Justus-von-Liebig-Weg 11, 37077 Goettingen, Germany; Fax: +49-551-39-25749; Email: ifeussn@uni-goettingen.de

Running Title: Lipid complement of *Arabidopsis* leaf mitochondria

Keywords: *Arabidopsis thaliana*, lipidome, lipid metabolism, lipid trafficking, mitochondria.

Summary

Mitochondria are considered as power stations of the cell, playing critical roles in various biological processes such as energy conversion, stress responses and programmed cell death. To maintain the structural and functional integrities of mitochondria, it is crucial to achieve a defined membrane lipid composition between different lipid classes wherein specific proportions of individual lipid species are present. Although mitochondria are capable of self-synthesizing a few lipid classes, many phospholipids are synthesized in the endoplasmic reticulum and transferred to mitochondria via membrane contact sites, as mitochondria are excluded from the vesicular transportation pathway. However, knowledge on the capability of lipid biosynthesis in mitochondria and the precise mechanism of maintaining the homeostasis of mitochondrial lipids is still scarce. Here we describe the lipidome of mitochondria isolated from *Arabidopsis* leaves, including the molecular species of glycerolipids, sphingolipids and sterols to depict the lipid landscape of mitochondrial membranes. In addition, we define proteins involved in lipid metabolism by proteomic analysis and apply them to reconstruct the lipid biosynthesis pathways in this organelle. Proteins putatively localized to the membrane contact sites are proposed based on the proteomic results and online databases. Collectively, our results suggest that leaf mitochondria are capable, with the assistant of membrane contact site-localized proteins, in generating several lipid classes including phosphatidylethanolamines, phosphatidic acids, phosphatidylinositols, phosphoinositides, phosphatidylglycerols, cardiolipins, diacylgalactosylglycerols and free sterols. We anticipate our work to be a foundation to further investigate the functional roles of lipids and their involvement in biochemical reactions in plant mitochondria.

Significance statement

Achieving and maintaining the specific membrane lipid composition is essential for plant mitochondria to stabilize their ultrastructure, and to keep their functional states. A combinatorial approach of lipidomics, proteomics and online database mining was applied to profile the mitochondrial lipidome in detail, depict the lipid landscape of plant mitochondria and assess their capability for lipid biosynthesis and modification.

Introduction

Mitochondria are considered as semiautonomous organelles. According to a widely accepted hypothesis, they descend from proteobacteria that have been engulfed by a eukaryotic cell. The two involved cells established an endosymbiosis (Gray *et al.* 1999). Mitochondria play crucial roles in various cellular processes, including ATP generation, stress responses and initiation of programmed cell death (Jacoby *et al.* 2012). The mitochondria of plant cells even have extended functions, many of which are related to photosynthesis (Braun 2020). Mitochondria are enclosed by two membranes, the outer (OM) and the inner (IM) mitochondrial membranes, that both require defined protein and lipid compositions to maintain their functional integrity (van Meer *et al.* 2008). The majority of the mitochondrial proteins is encoded by the nuclear genome, synthesized in the cytoplasm and post-translationally imported into the organelle. At the same time, a few proteins are encoded by the mitochondrial genome and synthesized within the organelle. Recent studies have largely broadened the knowledge of protein import in mitochondria, to the OM, the intermembrane space, the IM and the matrix (Wiedemann and Pfanner 2017). However, current understanding of the biosynthesis and transport of lipids, although fundamentally important constituents of the mitochondria, is still very limited. There are two major routes of lipid transport in the cell, vesicular and non-vesicular lipid transport. Vesicular lipid transport may be the predominant mechanism to deliver lipids from the site of synthesis to the final destination. Endoplasmic reticulum (ER), Golgi apparatus, plasma membrane and vacuoles use this pathway to exchange macromolecules between the compartments (Paul *et al.* 2014). However, mitochondria are disconnected from this vesicular transport and use primarily the non-vesicular pathway (Mesmin 2016). Non-vesicular transport is either mediated by lipid transfer proteins or direct lipid transport between two membranes. While the non-vesicular transport plays a key role in sorting certain lipids and assisting interorganellar communication, knowledge about its mechanism is still in its infancy (Lev 2010, Levine 2004).

Wide varieties of lipid species compose biological membranes. According to the backbones of the lipid molecules, they are classified into three main lipid categories – glycerolipids, sphingolipids and sterols. To delineate the molecular species in this study, we later use the taxonomy containing two colon-separated units for the numbers of carbons and double bonds in the fatty acyl moiety, and an additional unit after a semicolon for the number of hydroxyl groups (when present). Mitochondrial membranes contain high amounts of glycerophospholipids, such as phosphatidylcholine (PC) and phosphatidylethanolamine (PE), but are low in sphingolipids and sterols (Daum and Vance 1997). Most of these lipids are produced in the ER and are sorted to mitochondria and other organelles afterwards. Nevertheless, mitochondria have also their own capacity to generate some specific lipid classes. For instance, cardiolipin (CL) is synthesized in the IM and is present exclusively in mitochondria. It plays essential roles in establishing the cristae in the IM and in maintaining the mitochondrial ultrastructure. CL is formed through the condensation of phosphatidylglycerol (PG) and diacylglycerol (DAG) and finally consists of four acyl chains. The ER is the major site for generating PG molecules (Li-Beisson *et al.* 2013). However, enzymes involved in PG biosynthesis are identified in mitochondria as well, suggesting that mitochondria are capable of self-synthesizing PG and thus CL. In plants, the PG synthesizing enzymes, phosphatidylglycerolphosphate (PGP) synthase and PGP phosphatase (PGPP), are associated with mitochondria, ER and plastids; whereas CL synthase (CLS) localizes exclusively in the IM (Katayama *et al.* 2004, Nowicki *et al.* 2005, Xu *et al.* 2002). Notably, although mitochondrial lipid biosynthesis is not the major lipid source in plant cells, it is critical for certain organisms. In yeast, mitochondria are the major supplier of PE. They generate PE from phosphatidylserine (PS) by phosphatidylserine decarboxylase (PSD). In *Arabidopsis*, three PSD enzymes have been identified and PSD1 localizes in mitochondria, providing PE molecules *in situ* (Nerlich *et al.* 2007). Reversely, some PS molecules are converted from PE through the base-exchange pathway that substitutes the head groups of PE with serine molecules by PS synthase1, PSS1 (Yamaoka *et al.* 2011). While knowing mitochondria are capable of synthesizing PG, CL, PS and PE, its competence toward other lipid classes is still poorly understood (Li-Beisson *et al.* 2013).

Lipid trafficking between ER, mitochondria and plastids is essential for mitochondrial membrane biogenesis. Based on studies in yeast and mammals, glycerophospholipid biosynthesis takes place at a distinct membrane stretch of the ER, the mitochondria-associated membrane (MAM), wherein both ER and mitochondrial proteins have been identified (Vance 1990). Elevated activities of PE, PC and PS synthesizing enzymes have been detected in purified yeast MAM. Moreover, this distinct membrane domain seems to occur ubiquitously among plants and animal cells, suggesting its critical role during evolution and in mediating lipid transfer between ER and mitochondria (Achleitner *et al.* 1999, Morré *et al.* 1971, Staehelin 1997). Although mitochondria and plastids work closely together in numerous pathways in plants, the lipid transport mechanism between these two organelles is largely unknown. Nevertheless, a few studies have suggested the relevance of lipid trafficking between these two organelles for survival, especially under environmental stresses. During phosphate starvation, higher numbers of mitochondria – plastid junctions are established (Jouhet *et al.* 2004). At the same time, drastic lipid remodeling of mitochondria, plastids and plasma membrane arises. PC and PE are degraded to release the phosphate residues for essential biological processes and the remaining molecules are recycled to generate the typical plastidial glyceroglycolipid, digalactosyldiacylglycerol (DGDG). This coincides with increased levels of CL in mitochondria isolated from *Arabidopsis* suspension cells and calli. In *Arabidopsis*, the protein complex involved in lipid trafficking and tethering of the two mitochondrial membranes, the mitochondrial transmembrane lipoprotein (MTL) complex, has been identified recently (Michaud *et al.* 2016). MTL is composed of more than 200 subunits and it has been demonstrated that MTL promotes the translocation of PE from IM to OM and the import of DGDG from plastids to mitochondria during phosphate starvation. Furthermore, mutation of a newly identified MTL subunit, digalactosyldiacylglycerol synthase suppressor 1 (DGS1), leads to alteration of plastidial and mitochondrial lipid composition and deficiency in mitochondrial biogenesis (Li *et al.* 2019). The characterization of the MTL complex provides an initial insight in understanding the mechanism of lipid trafficking between mitochondria and plastids.

In this study, we aimed to characterize the lipid metabolism of *Arabidopsis* leaf mitochondria and conducted an in-depth lipidomic analysis, providing the molecular species information of all lipid categories including glycerolipids, sphingolipids and sterols to illustrate the lipid landscape of mitochondria in *Arabidopsis* leaves. In combination with a proteomic approach, we intended to specify the capacity of lipid biosynthesis and modification in plant mitochondria, defining lipid species and classes that may be generated by mitochondria. We additionally propose putative membrane contact site-localizing proteins and their roles in interorganellar communication. Our results suggest that leaf mitochondria possess a defined lipid composition wherein specific lipid molecular species appear. In addition, lipid trafficking between mitochondria, ER and plastids via membrane contact sites provides assistance in maintaining the homeostasis of the lipid composition in mitochondria.

Results

Purity of mitochondrial fractions

A combinatorial approach by lipidomics, proteomics and mining of online databases was applied to investigate the lipid composition as well as the capacity of lipid metabolism of *Arabidopsis* leaf mitochondria. Mitochondria were isolated by differential centrifugation combined with Percoll density gradient centrifugation. To ensure and evaluate the purity of the mitochondrial fractions, all samples were investigated by two-dimensional (2D) blue native (BN)/SDS polyacrylamide gel electrophoresis (PAGE) and by shotgun proteome analyses. Our mitochondrial fractions proved to be of good quality. The two photosystems were not detectable on our 2D gels. Only trace amounts of the large subunit from Rubisco, the most abundant protein in leaves, were visible in the leaf mitochondrial fractions (L-mito) on BN/SDS gels, and the small subunit of Rubisco was barely visible (Fig. 1A, Fig. S1). Mitochondria isolated from dark-grown *Arabidopsis* cell culture (C-mito) were investigated by the same procedure and used as a negative control for contamination from chloroplasts. The L-mito and C-mito fractions were highly similar (Fig. 1, Fig. S1). In addition, the purity of the mitochondrial fractions was investigated by label-free quantitative shotgun proteomics. Building on the proteome data (Sup. Tab. 1), the proportion of mitochondria-localized proteins was calculated to be in the range of 87 to 94 % (Fig. 1B, Fig. S2). These results indicate that the mitochondrial fractions are of high purity. The following sections compare the lipidome and proteome of L-mito with total leaf extracts (L-TE); the C-mito samples were used as a quality control.

PC, PE and CL are strongly enriched glycerolipids in plant leaf mitochondria

Glycerolipids are the most abundant lipids not only in mitochondria but also generally in plant tissues, accounting for more than 90 % of the overall lipids in leaves (Li-Beisson *et al.* 2013). Glycerophospholipids are fundamental to all biological membranes, while glyceroglycolipids are critical for photosynthetic membranes and localize mainly in plastids of vegetative tissues (Hölzl and Dörmann 2019). Therefore, a quantitative approach based on TLC-GC was applied on determining the proportion of the major glycerolipid classes in L-TE and L-mito (Fig. 2). In L-TE, glyceroglycolipids including monogalactosyldiacylglycerols (MGDG; 38.9 ± 1.1 %) and DGDG (22.3 ± 3.8 %) contribute to the majority of the overall lipids; while glycerophospholipids, PC (35.7 ± 6.0 %) and PE (37.2 ± 0.3 %), are the most abundant lipids in L-mito. CL is an essential component for establishing the mitochondrial cristae in IM. It accounts for 4.0 ± 0.9 % in the lipidome of L-mito; however, it was not detectable in L-TE because of its low abundance in total cellular lipids. PG, as the precursor of CL, consists 5.6 ± 2.1 % of the overall lipids in L-mito.

Specific molecular glycerolipid species are enriched in plant leaf mitochondria

Glycerophospholipids are the most abundant lipids in mitochondria (Fig. 2) (Jouhet *et al.* 2004, Michaud *et al.* 2016). The most abundant PC and PE molecular species in mitochondria with the acyl chains 18:2/18:3, 18:2/18:2 and 18:3/18:3 account for more than 60 % in both lipid classes (Fig. 3a; Tab. S2). The species 16:0/18:2, 16:0/18:3 of both lipid classes however were depleted by about 50 % in this organelle in comparison to L-TE. PS (18:0/18:2) composes up to 38.3 % in L-mito with an increase of $2^{4.3}$ fold comparing to L-TE. The accumulation of this lipid species was the most remarkable one that was even visible when the fatty acid profiles of the different lipid classes were calculated from the molecular species data, showing an enrichment of 18:0 in mitochondria on the expense of 16:0 (Fig. S3). Together the main molecular species of these three lipid classes have an acyl chain composition (C18/C18) in common, which suggests a common origin.

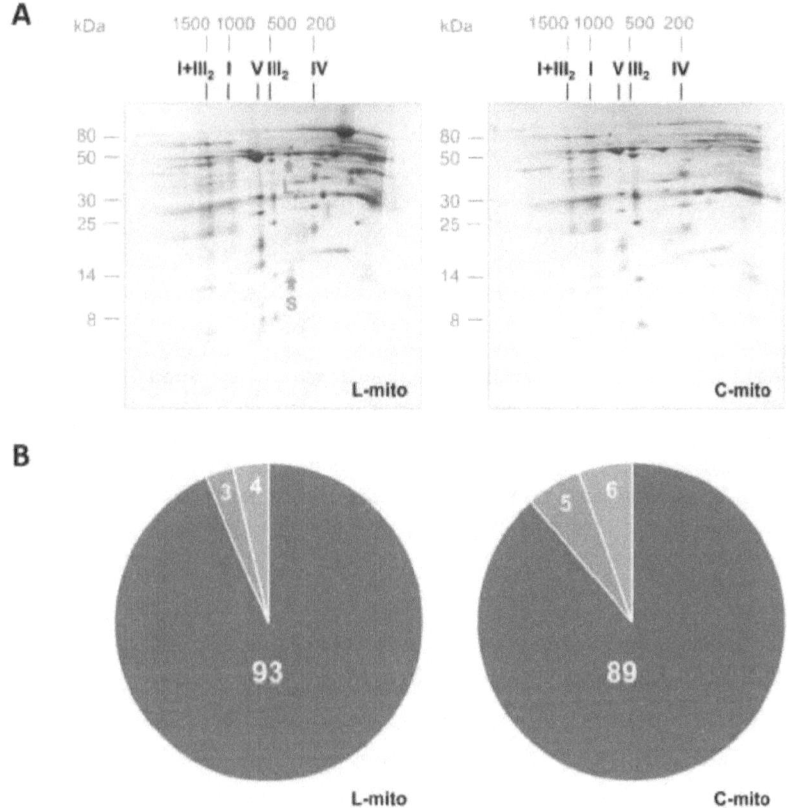

Figure 1. Purity of mitochondrial fractions.
The purity of mitochondrial fractions was determined by 2D blue native / SDS PAGE **(A)** and by summed-up peptide intensities of subcellular compartments based on protein assignments as given by the Subcellular localization database for Arabidopsis proteins (SUBAcon; www.suba.live) **(B)**. **A**: Mitochondria were isolated from *Arabidopsis* leaves (L-mito) and cell cultures (C-mito). Proteins were separated by 2D Blue native PAGE and Coomassie-stained. Numbers on top and to the left of the 2D gels refer to the masses of standard protein complexes / proteins (in kDa), the roman numbers above the gels to the identity of OXPHOS complexes. I+III₂: supercomplex consisting of complex I and dimeric complex III; I: complex I; V: complex V; III₂: dimeric complex III; IV: complex IV. The small (S; 14.5 kDa) and the large (L; 53.5 kDa) subunit of Rubisco are indicated by green arrows. Biological replicates of the 2D gels and reference gels for mitochondrial and chloroplast fractions from *Arabidopsis* are shown in Fig. S1. **B**: Mitochondrial fractions from *Arabidopsis* leaves and cell cultures were analyzed by label-free quantitative shotgun proteomics. Peptide intensities assigned to subcellular compartments were summed-up and averaged results for L-mitos and C-mitos were visualized by pie charts (for detailed results see Fig. S2). Blue: mitochondria; green: plastids; gray: others; numbers in %.

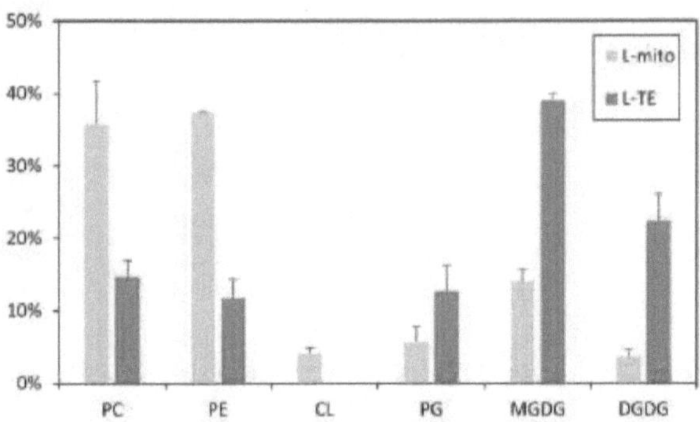

Figure 2. Lipid class profiles of purified mitochondria and total leaf extracts.
Glycerolipids of leaf total extract (L-TE) and mitochondria isolated from leaves (L-mito) were analyzed quantitatively by TLC-GC approach. Data of L-TE represent mean values in mol % from three independent experiments ± SD; data of L-mito represent mean values in mol % from two independent experiments ± SEM. PC, phosphatidylcholine; PE, phosphatidylethanolamine; CL, cardiolipin; PG, phosphatidylglycerol; MGDG, monogalactosyldiacylglycerol; DGDG, digalactosyldiacylglycerol.

An important intermediate in glycerolipid metabolism is phosphatidic acid (PA), serving as precursor or product for numerous glycerolipids including PC, PE, PG, phosphatidylinositol (PI) and DAG. PA (16:0/18:2), PA (16:0/18:3) and PA (18:2/18:3) are the most abundant PA species in both fractions, L-mito and L-TE. DAG as the other important lipid intermediate displays a higher complexity in its species profile. DAG (16:0/18:2), DAG (16:0/18:3), DAG (18:2/18:3) and DAG (18:3/18:3) build a substantial amount (>65 mol %) of the DAG profile in L-mito (Fig. 3c).

PG is formed from PA and serves as precursor for the mitochondrial lipid CL. Hydrolysis and condensation of PG lead to the formation of DAG and CL correspondingly. PG (16:0/18:2) and PG (16:0/18:3) are the major species in L-mito, but only the former species is enriched in mitochondria ($2^{1.1}$ fold). In addition, unsaturated PG species are significantly enriched in L-mito, which contribute to the structures of the down-stream CL molecules. CL (72:10) and CL (72:11), composed of 18:2 and 18:3 acyl chains, are the most abundant CL (Fig. 3a).

Glycerophospholipids serve not only as membrane building blocks but are also important precursors for signaling molecules in the cell. PI, phosphatidylinositol monophosphate (PIP) and phosphatidylinositol bisphosphate (PIP$_2$) exert regulatory functions in cell development and polarity determination (Heilmann 2016). In L-mito, molecular species with 16:0/18:2 and 16:0/18:3 acyl chains compose up to 80 % of PI and PIP$_2$; yet PIP was not detectable in L-mito. PI (16:0/16:1) and PI (18:0/18:1) were only detectable in the mitochondrial extracts, and PI (18:0/18:2) is significantly enriched in L-mito for $2^{1.2}$ fold in comparison to L-TE (Fig. 3a). On the other hand, we obtained higher signals of PIP$_2$ (18:1/18:2) in L-mito.

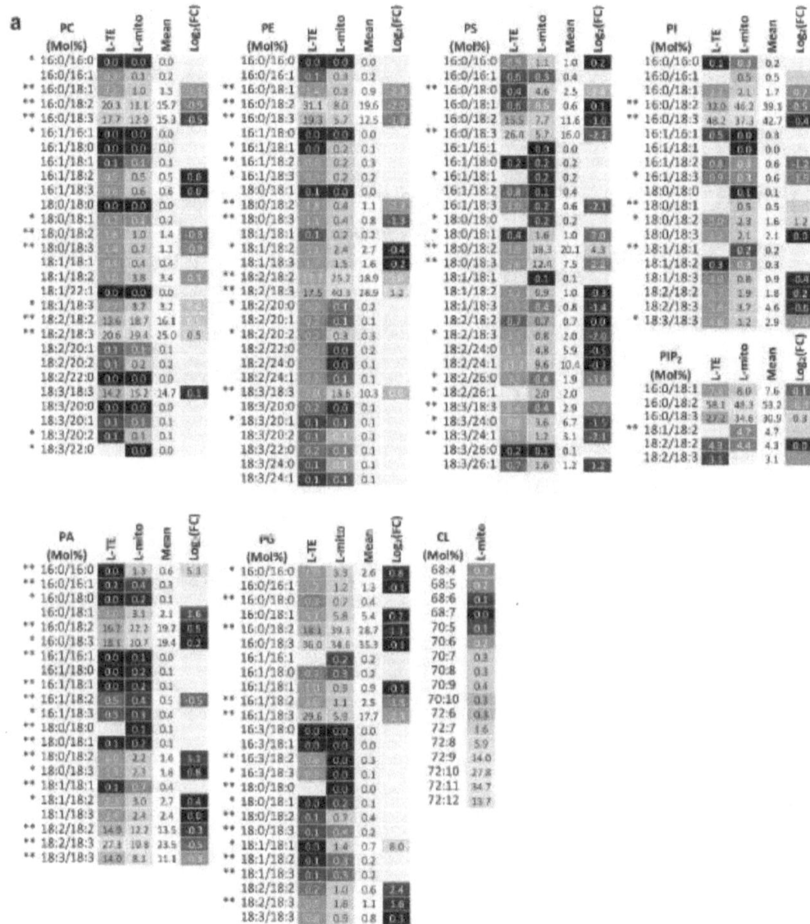

b

MGDG (Mol%)

		L-TE	L-mito	Mean	Log$_2$(FC)
*	16:0/16:0	0.0	0.0	0.0	
	16:0/16:1	0.0	0.0	0.0	
*	16:0/16:2	0.0	0.1	0.0	
**	16:0/16:3	0.1	0.1	0.1	
	16:0/18:0	0.0	0.1	0.1	
**	16:0/18:1	0.6	1.4	1.0	1.1
	16:0/18:2	0.5	0.9	0.7	0.7
*	16:0/18:3	0.8	2.3	1.6	1.5
	16:1/16:1	0.0	0.0	0.0	
	16:1/16:2	0.0	0.0	0.0	
	16:1/16:3	0.1	0.1	0.1	
	16:1/18:0	0.1	0.1	0.1	
*	16:1/18:1	0.8	0.5	0.7	-0.8
	16:1/18:2	1.1	1.0	1.0	-0.1
*	16:1/18:3	1.3	0.9	1.1	-0.5
	16:2/18:0	0.0	0.0	0.0	
	16:2/18:1	0.4	0.4	0.4	
	16:2/18:2	1.4	1.2	1.3	-0.2
**	16:2/18:3	3.6	2.6	3.1	-0.4
	16:3/18:0	0.0	0.0	0.0	
*	16:3/18:1	0.3	0.2	0.3	
**	16:3/18:2	1.6	1.1	1.3	-0.6
**	16:3/18:3	63.8	52.1	57.9	-0.3
**	18:0/18:0	0.0	0.0	0.0	
*	18:0/18:1	0.0	0.1	0.0	
**	18:0/18:2	0.0	0.2	0.1	
**	18:0/18:3	0.1	0.3	0.2	
	18:1/18:1	0.0	0.2	0.1	
	18:1/18:2	0.1	0.2	0.1	
*	18:1/18:3	0.5	0.8	0.6	0.6
*	18:2/18:2	0.5	1.2	0.9	1.4
**	18:2/18:3	2.5	4.2	3.4	0.7
*	18:3/18:3	19.6	27.7	23.7	0.5

DGDG (Mol%)

		L-TE	L-mito	Mean	Log$_2$(FC)
**	16:0/16:0	0.1	0.3	0.2	
	16:0/16:1	0.0	0.0	0.0	
	16:0/16:2	0.0	0.1	0.0	
	16:0/16:3	0.0	0.1	0.0	
**	16:0/18:0	0.2	0.1	0.1	
	16:0/18:1	1.3	1.2	1.2	-0.1
*	16:0/18:2	5.3	3.7	4.5	-0.5
	16:0/18:3	11.8	11.4	11.6	-0.1
	16:1/16:1		0.0	0.0	
	16:1/16:2		0.0	0.0	
	16:1/16:3		0.0	0.0	
	16:1/18:0	0.0	0.0	0.0	
	16:1/18:1	0.1	0.1	0.1	
	16:1/18:2	0.2	0.2	0.2	
	16:1/18:3	0.5	0.5	0.5	-0.1
	16:2/18:0	0.0	0.1	0.0	
	16:2/18:1	0.1	0.1	0.1	
	16:2/18:2	0.2	0.3	0.3	
	16:2/18:3	1.9	1.4	1.7	-0.4
	16:3/18:0	0.0	0.1	0.1	
	16:3/18:1	0.0	0.1	0.0	
*	16:3/18:2	0.2	0.4	0.3	
*	16:3/18:3	10.1	14.0	12.0	0.5
	18:0/18:0		0.0	0.0	
**	18:0/18:1	0.0	0.1	0.1	
	18:0/18:2	0.2	0.2	0.2	
	18:0/18:3	0.8	1.0	0.9	0.2
	18:1/18:1	0.0	0.0	0.0	
	18:1/18:2	0.1	0.1	0.1	
**	18:1/18:3	0.8	1.0	0.9	0.3
	18:2/18:2	0.4	0.4	0.4	
	18:2/18:3	3.2	3.1	3.1	0.0
	18:3/18:3	62.2	60.0	61.1	-0.1

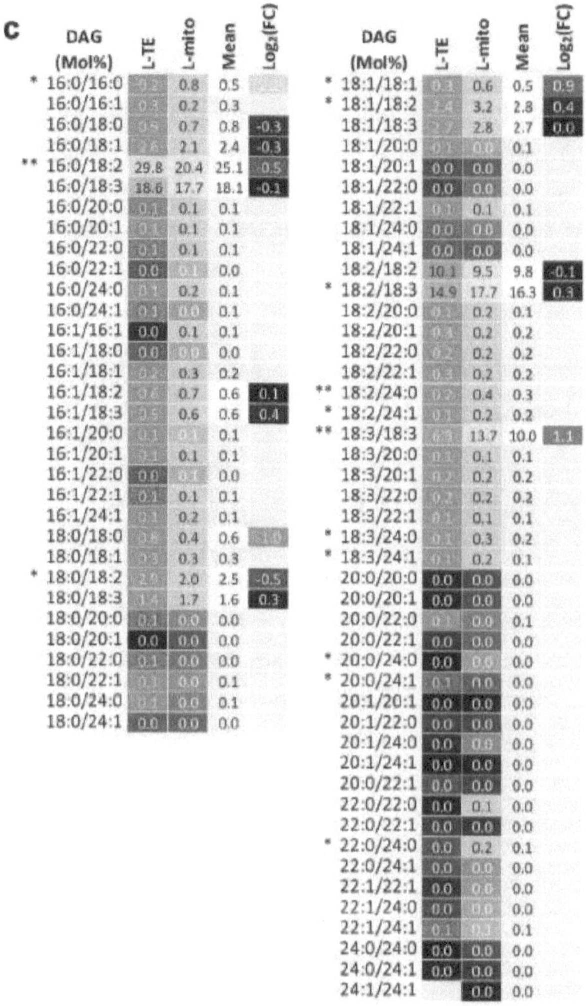

c

	DAG (Mol%)	L-TE	L-mito	Mean	Log$_2$(FC)
*	16:0/16:0	0.2	0.8	0.5	
	16:0/16:1	0.3	0.2	0.3	
	16:0/18:0	0.9	0.7	0.8	-0.3
	16:0/18:1	2.6	2.1	2.4	-0.3
**	16:0/18:2	29.8	20.4	25.1	0.5
	16:0/18:3	18.6	17.7	18.1	-0.1
	16:0/20:0	0.1	0.1	0.1	
	16:0/20:1	0.1	0.1	0.1	
	16:0/22:0	0.1	0.1	0.1	
	16:0/22:1	0.0	0.1	0.0	
	16:0/24:0	0.1	0.2	0.1	
	16:0/24:1	0.1	0.0	0.1	
	16:1/16:1	0.0	0.1	0.1	
	16:1/18:0	0.0	0.0	0.0	
	16:1/18:1	0.2	0.3	0.2	
	16:1/18:2	0.6	0.7	0.6	0.1
	16:1/18:3	0.5	0.6	0.6	0.4
	16:1/20:0	0.1	0.1	0.1	
	16:1/20:1	0.1	0.1	0.1	
	16:1/22:0	0.0	0.1	0.0	
	16:1/22:1	0.1	0.1	0.1	
	16:1/24:1	0.1	0.2	0.1	
	18:0/18:0	0.8	0.4	0.6	1.0
	18:0/18:1	0.3	0.3	0.3	
*	18:0/18:2	2.9	2.0	2.5	-0.5
	18:0/18:3	1.4	1.7	1.6	0.3
	18:0/20:0	0.1	0.0	0.0	
	18:0/20:1	0.0	0.0	0.0	
	18:0/22:0	0.1	0.0	0.0	
	18:0/22:1	0.1	0.0	0.1	
	18:0/24:0	0.1	0.0	0.1	
	18:0/24:1	0.0	0.0	0.0	

	DAG (Mol%)	L-TE	L-mito	Mean	Log$_2$(FC)
*	18:1/18:1	0.3	0.6	0.5	0.9
*	18:1/18:2	2.4	3.2	2.8	0.4
	18:1/18:3	2.7	2.8	2.7	0.0
	18:1/20:0	0.1	0.0	0.1	
	18:1/20:1	0.0	0.0	0.0	
	18:1/22:0	0.0	0.0	0.0	
	18:1/22:1	0.1	0.1	0.1	
	18:1/24:0	0.0	0.0	0.0	
	18:1/24:1	0.0	0.0	0.0	
	18:2/18:2	10.1	9.5	9.8	-0.1
*	18:2/18:3	14.9	17.7	16.3	0.3
	18:2/20:0	0.1	0.2	0.1	
	18:2/20:1	0.1	0.2	0.2	
	18:2/22:0	0.2	0.2	0.2	
	18:2/22:1	0.3	0.2	0.2	
**	18:2/24:0	0.2	0.4	0.3	
*	18:2/24:1	0.1	0.2	0.2	
**	18:3/18:3	6.3	13.7	10.0	1.1
	18:3/20:0	0.1	0.1	0.1	
	18:3/20:1	0.2	0.2	0.2	
	18:3/22:0	0.2	0.2	0.2	
	18:3/22:1	0.1	0.1	0.1	
*	18:3/24:0	0.1	0.3	0.2	
*	18:3/24:1	0.1	0.2	0.1	
	20:0/20:0	0.0	0.0	0.0	
	20:0/20:1	0.0	0.0	0.0	
	20:0/22:0	0.1	0.0	0.1	
	20:0/22:1	0.0	0.0	0.0	
*	20:0/24:0	0.0	0.0	0.0	
*	20:0/24:1	0.1	0.0	0.0	
	20:1/20:1	0.0	0.0	0.0	
	20:1/22:0	0.0	0.0	0.0	
	20:1/24:0	0.0	0.0	0.0	
	20:1/24:1	0.0	0.0	0.0	
	20:0/22:1	0.0	0.0	0.0	
	22:0/22:0	0.0	0.1	0.0	
	22:0/22:1	0.0	0.0	0.0	
*	22:0/24:0	0.0	0.2	0.1	
	22:0/24:1	0.0	0.0	0.0	
	22:1/22:1	0.0	0.0	0.0	
	22:1/24:0	0.0	0.0	0.0	
	22:1/24:1	0.1	0.1	0.1	
	24:0/24:0	0.0	0.0	0.0	
	24:0/24:1	0.0	0.0	0.0	
	24:1/24:1		0.0	0.0	

Figure 3. Profiles of molecular glycerolipid species and the according fold changes between L-mito and L-TE.

Heat maps of (a) glycerophospholipids, (b) glyceroglycolipids and (c) diacylglycerols illustrate the difference of species distribution based on LC-MS/MS analyses. Each block represents one lipid class wherein the detected species are presented and summarized to 100 %. Data represent mean values in mol % from three independent experiments. Binary logarithm was applied when the mean values are higher than 0.5 to inspect the fold changes (FC) between L-mito and L-TE. One and two asterisks (*, **) indicate p values < 0.05 and < 0.01, respectively, by Student's t-test. PC, phosphatidylcholine; PE, phosphatidylethanolamine; PS: phosphatidylserine; PI: phosphatidylinositol; PA phosphatidic acid; DAG: Diacylglycerol; CL, cardiolipin; PG, phosphatidylglycerol; MGDG, monogalactosyldiacylglycerol; DGDG, digalactosyldiacylglycerol.

Glyceroglycolipids carry carbohydrate residues as their head groups; for instance, the galactose-containing lipids, MGDG and DGDG contain one and two galactoses, respectively (Hölzl and Dörmann 2019). MGDG (16:3/18:3) and DGDG (18:3/18:3) are the major components of the overall glyceroglycolipids in both L-TE and L-mito (Fig. 3b). The amount of MGDG (16:0/18:3) and MGDG (18:2/18:2) are specifically elevated in L-mito for $2^{1.5}$ and $2^{1.4}$ folds, respectively, compared to L-TE. In contrast, the species MGDG (16:1/18:1) and MGDG (16:3/18:2) are significantly decreased in L-mito. In summary, the major mitochondrial lipids PC, PE, PS and CL consist of similar species, while PI, PIP2 and PG consist primarily of different species and the lipid precursors DAG and PA resemble a mixture of both origins.

Combining lipidomic and proteomic analyses provides insights into the leaf mitochondrial glycerolipid metabolism capacity

To analyze the lipid metabolism capacity of the mitochondria, their protein composition and abundance was investigated via label-free quantitative shotgun proteomics. Subsequently, their biological function and the subcellular localization were retrieved from The Arabidopsis Information Resource (TAIR) and the Subcellular localization database for Arabidopsis proteins (SUBAcon) databases (Fig. S4) (Hooper *et al.* 2017, Lamesch *et al.* 2011). About 40 proteins involved in the biosynthesis and modification of fatty acids and more complex lipids were identified in the mitochondrial extracts (Tab. S3). With the assistance of the Kyoto Encyclopedia of Genes and Genomes (KEGG) (Tanabe and Kanehisa 2012), this list was used to annotate mitochondrial lipid biosynthesis pathways including glycerolipid, sphingolipid and sterol metabolism, which were then compared with the identified mitochondrial molecular lipid species in the following sections.

More than 60 mol % of both PC and PE species carry two C18 fatty acids on their glycerol backbone. This strongly suggests that they may be specifically imported from the ER through ER – mitochondrial contact sites (Michaud *et al.* 2016). Both lipid classes can be interconverted into each other via PA and the detected PLDα1 enzyme activity (Fig. 4a). In addition, molecular PE species may be products of two mitochondrial pathways: (i) via PSD1 which was identified in all mitochondrial samples with higher abundance in C-mito and (ii) via the cytidine diphosphate (CDP)-ethanolamine pathway (Nerlich *et al.* 2007). It comprises a series of reactions catalyzed by ethanolamine kinase, ethanolamine phosphate cytidylyltransferase (PECT) and aminoalcohol phosphotransferase (AAPT). PECT is localized at mitochondrial periphery in *Arabidopsis* (Mizoi *et al.* 2006), and AAPT at ER and Golgi apparatus in wheat (Sutoh *et al.* 2010). Indeed the former finding was confirmed in our study by similar abundances of PECT1 in the L-mito and C-mito samples. PS can be synthesized from PE via base-exchange reactions conducted by PS synthase 1 (PSS1) in mitochondria (Yamaoka *et al.* 2011). However, we were not able to detect this protein in any of our samples.

DAG serves as a central hinge in glycerolipid metabolism. It can be converted from or to numerous glycerophospholipid classes, such as PC, PE and PA (Fig. 4a). PG is synthesized from PA via CDP-DAG by PGP1. Next, CL is synthesized after the condensation of PG and CDP-DAG by CL synthase (CLS) or after the transacylation of monolyso-CL by acyltransferases such as LCLAT (Tafazzin) (Xu *et al.* 2006). CLS and LCLAT were identified in the shotgun proteomic analysis in this study with a higher abundance in C-mito compared to L-mito.

PIP and PIP$_2$ are products of PI kinases, for instance FAB1B, and can be degraded by phospholipase C (PLC), inositol polyphosphate phosphatase (SAL1) and/or CDP-DAG synthases (Fig. 4c). In *Arabidopsis*, PI is synthesized by PI synthase that utilizes CDP-DAG and *myo*-inositol as substrates. In mammalian cells, PI biosynthesis takes place mainly in ER. However, it has been challenged recently by a novel hypothesis that PI synthase is located at a highly mobile membrane, which synthesizes and delivers phosphoinositides to other intracellular compartments (Kim *et al.* 2011). Although many enzymes involved in phosphoinositide metabolism have been annotated to localize in mitochondria, only SAL1 was identified in our approach.

The biosynthesis of glyceroglycolipids takes place in plastids by MGDG and DGDG synthases (MGDs and DGDs), respectively (Hölzl and Dörmann 2019). MGDG is generated by adding a galactose head group to DAG; subsequent addition of another galactose by DGD1 generates DGDG. The detected specific species profiles of mitochondrial glyceroglycolipids may suggest at least for a specific transport from plastids to mitochondria (Fig. 4b). The sulfolipid SQDG is synthesized in two steps. Uridine diphosphate (UDP)-glucose is first combined with a sulfite by UDP-sulfoquinovose synthase 1 (SQD1), followed by transferring the sulfoquinovose head group to DAG and thus forms SQDG. Although SQD1 was identified in the mitochondrial extracts in this study, we did not detect SQDG lipids in any of the mitochondrial samples. Together we could exclusively localize three biosynthetic steps to mitochondria: (i) CLS catalyzing the formation of CL, (ii) PSD1 transforming PS into PE and (iii) SAL1 which degrades IP$_2$.

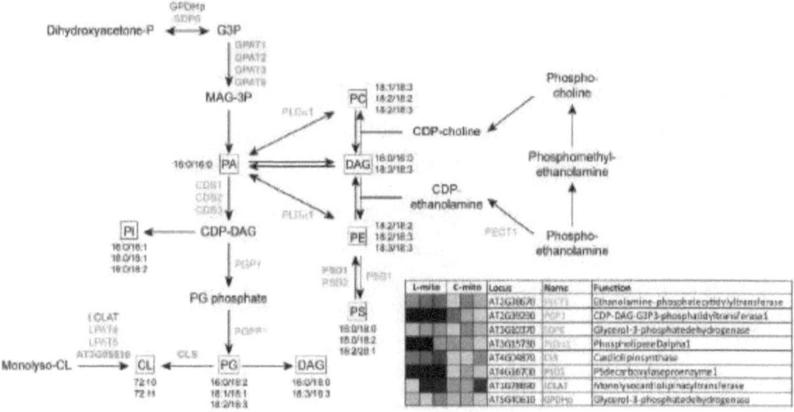

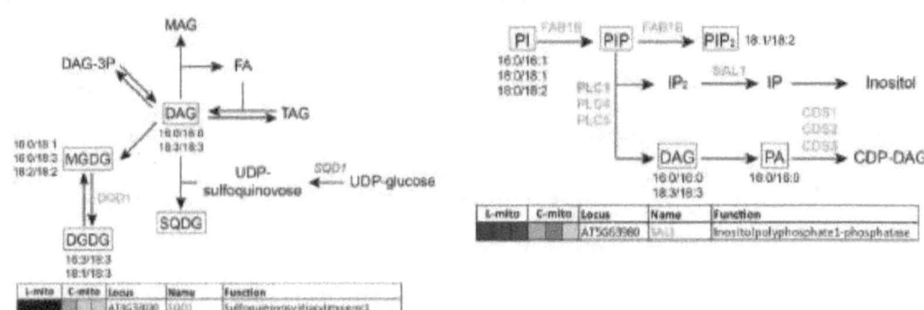

Figure 4. Mitochondrial localized proteins and enriched lipids related to general glycerolipid biosynthesis.

Pathways of (a) glycerophospholipid, (b) glyceroglycolipid and (c) phosphoinositide metabolism constitute the general glycerolipid biosynthesis in *Arabidopsis*. Proteins identified and/or localized in mitochondria were labeled (i) blue: proteins identified in the proteomic analysis of this study and also predicted to localize in mitochondria, (ii) green: proteins identified in the proteomic analysis of this study in mitochondria but predicted to localize in other organelles, (iii) magenta: proteins absent in the proteomic analysis of this study but predicted to localize or found in mitochondria, (iv) bold font: exclusively localized in mitochondria and (v) italic font: only identified in one of the mitochondrial populations. Heat maps visualize the protein abundance of three independent experiments of mitochondria purified from leaves and cell cultures. Boxed lipids: lipid classes analyzed by the LC-MS/MS approach with molecular lipid species enriched in L-mito in comparison to L-TE. Further full names and functions are itemized in Table S1. In the heat maps, proteins with high and low abundance are depicted in yellow and black, respectively. Predicted protein localization was based on The Arabidopsis Information Resource (TAIR; www.arabidopsis.org) and the Subcellular localization database for Arabidopsis proteins (SUBAcon; www.suba.live).

Specific molecular sphingolipid and sterol species suggest a function in mitochondria

Sphingolipids and sterols are important modulators of membrane microdomains and play critical roles in regulating the balance between cell survival and apoptosis. We profiled the mitochondrial sphingolipid compositions of both simple and complex sphingolipid groups, including long-chain bases (LCB), phosphorylated LCB (LCB-P), ceramides (Cer), glucosylceramides (GlcCer) and glycosyl inositol phosphoceramides (GIPC) (Fig. 5). All sphingolipids have LCBs, 18-carbon amino-alcohols, as their backbones. Phytosphingosine (18:0;3), hosting three hydroxyl groups, is the most abundant free LCB in both L-mito and L-TE (Fig. 5a). However, dihydrosphingosine (18:0;2) is highly enriched in L-mito as well as phosphorylated dihydrosphingosine (18:0;2-P), which is the major component in the LCB-P pool of L-mito (64.7 %). LCBs can be further N-acylated to generate Cer, the basic structure of complex sphingolipid classes. Addition of glucoses to Cer generates GlcCer, following sequential extension of phosphoinositol, hexose and/or hexose derivatives generate series of GIPCs. Series 0, A and B GIPCs carry one, two and three additional hexoses on the head groups of inositol phosphoceramides, respectively (Cacas *et al.* 2012). Similar to LCB and LCB-P, only specific species were enriched in L-mito (Fig. 5b): Cer (18:0;3/24:0;1), GlcCer (18:1;2/24:1;1) and series A hexose-carrying GIPC (H-GIPC) (18:1;3/24:1;1). Remarkably, GIPCs were only detectable in mitochondrial samples (both L-mito and C-mito) in our approach. However, proteins related to sphingolipid metabolism were neither identified in plant mitochondria via proteomic analyses, nor retrieved from online databases.

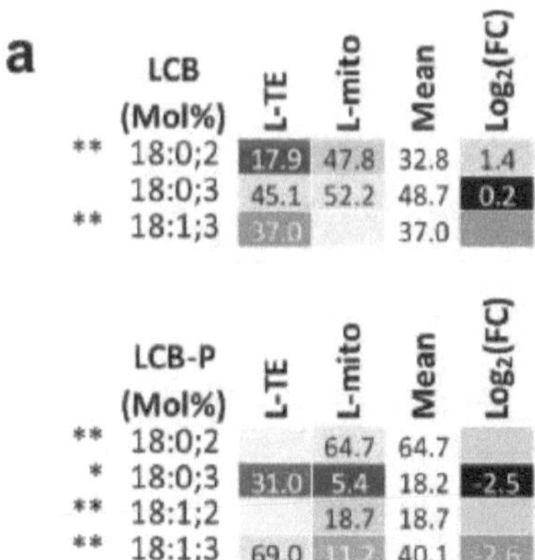

b

	Cer (Mol%)	L-TE	L-mito	Mean	Log₂(FC)
	18:0;2/16:0;0	0.1	0.5	0.3	
	18:0;2/16:0;1	0.9	0.6	0.7	-0.5
	18:0;3/16:0;0	0.9	0.4	0.7	-1.2
	18:0;3/16:0;1	0.7	0.8	0.7	0.2
*	18:0;3/22:0;0	1.6	0.6	1.1	-1.4
	18:0;3/22:0;1	1.5	0.8	1.2	-0.9
	18:0;3/22:1;0	0.2	1.5	0.9	3.0
**	18:0;3/22:1;1		0.3	0.3	
*	18:0;3/24:0;0	7.4	2.5	5.0	-1.5
**	18:0;3/24:0;1	3.1	16.3	9.7	2.4
**	18:0;3/24:1;0	1.2	4.0	2.6	1.7
	18:0;3/24:1;1	2.1	1.1	1.6	-0.9
	18:0;3/26:0;0	2.1	1.6	1.8	-0.4
**	18:0;3/26:0;1	1.4	4.7	3.1	1.7
	18:0;3/26:1;0	0.4	0.4	0.4	
	18:0;3/26:1;1	0.1	0.3	0.2	
**	18:1;2/16:0;0	0.2	3.1	1.6	4.0
	18:1;2/16:0;1	0.5	0.5	0.5	0.0
	18:1;2/24:0;0	1.7	1.5	1.6	-0.2
*	18:1;2/24:0;1	1.1	11.4	6.3	3.3
	18:1;2/24:1;0	0.9	5.8	3.4	2.7
	18:1;2/24:1;1	1.2	2.5	1.9	1.0
	18:1;2/26:0;0	0.9	0.7	0.8	-0.3
	18:1;2/26:0;1	1.3	0.9	1.1	-0.5
	18:1;2/26:1;0	0.3	0.8	0.6	1.4
*	18:1;2/26:1;1	0.3	2.3	1.3	3.2
	18:1;3/16:0;0	0.3	0.5	0.4	
	18:1;3/16:0;1	0.6	1.0	0.8	0.7
	18:1;3/20:0;0	0.1	0.2	0.1	
	18:1;3/20:0;1	0.3	0.2	0.2	
	18:1;3/20:1;0	0.3	0.1	0.2	
	18:1;3/20:1;1	0.1	0.2	0.1	
**	18:1;3/22:0;0	2.7	0.5	1.6	-2.6
	18:1;3/22:0;1	2.7	1.4	2.0	-1.0
	18:1;3/22:1;0	0.1	0.5	0.3	
	18:1;3/22:1;1	0.1	0.1	0.1	
**	18:1;3/24:0;0	9.2	2.0	5.6	-2.2
*	18:1;3/24:0;1	4.3	5.9	5.1	0.5
**	18:1;3/24:1;0	7.0	1.3	4.1	-2.5
	18:1;3/24:1;1	7.6	5.4	6.5	-0.5
*	18:1;3/26:0;0	6.6	2.4	4.5	-1.5
*	18:1;3/26:0;1	6.5	3.3	4.9	-1.0
	18:1;3/26:1;0	1.9	0.9	1.4	-1.1
*	18:1;3/26:1;1	0.9	1.2	1.0	0.4
	18:1;3/28:0;0	0.9	0.6	0.8	-0.7
	18:1;3/28:0;1	0.7	0.7	0.7	0.2
**	18:1;3/28:1;0	1.0	0.1	0.6	-2.9
	18:1;3/28:1;1	0.1	0.5	0.3	
	18:2;2/16:0;0	0.2	0.2	0.2	
	18:2;2/16:0;1	0.3	0.3	0.3	
	18:2;2/24:0;0	0.5	0.4	0.5	-0.5
	18:2;2/24:0;1	8.0	1.7	4.9	-2.2
	18:2;2/24:1;0	0.8	0.2	0.5	-2.2
*	18:2;2/24:1;1	0.5	0.2	0.3	
	18:2;2/26:0;0	0.2	0.2	0.2	
*	18:2;2/26:0;1	2.9	1.2	2.1	-1.2
	18:2;2/26:1;0	0.3	0.2	0.2	
	18:2;2/26:1;1	0.2	0.1	0.2	

	GlcCer (Mol%)	L-TE	L-mito	Mean	Log₂(FC)
**	18:0;3/16:0;0	0.7	0.2	0.4	
	18:0;3/16:0;1	0.4	0.2	0.3	
	18:0;3/24:0;0	0.2	0.1	0.2	
**	18:0;3/24:0;1	0.6	0.2	0.4	
	18:0;3/24:1;0	0.2	0.6	0.4	
**	18:0;3/24:1;1	1.5	0.4	1.0	-2.0
	18:0;3/26:0;0	0.1	0.1	0.1	
	18:0;3/26:0;1	0.2	0.2	0.2	
	18:0;3/26:1;0	0.0	0.0	0.0	
	18:0;3/26:1;1	0.2	0.2	0.2	
	18:1;2/16:0;0	0.3	0.6	0.5	0.8
**	18:1;2/16:0;1		2.4	2.4	
	18:1;2/24:0;0	0.0	0.1	0.1	
**	18:1;2/24:0;1	0.2	4.4	2.3	[illegible]
**	18:1;2/24:1;0	0.1	0.6	0.3	
**	18:1;2/24:1;1	1.1	64.3	32.7	[illegible]
*	18:1;2/26:0;0	0.1	0.4	0.2	
*	18:1;2/26:0;1	0.2	0.7	0.4	
	18:1;2/26:1;0	0.1	0.2	0.1	
	18:1;2/26:1;1	0.2	0.2	0.2	
*	18:1;3/16:0;0	0.5	0.2	0.3	
**	18:1;3/16:0;1	14.3	4.3	9.3	-1.7
	18:1;3/16:1;0	0.0	0.0	0.0	
	18:1;3/16:1;1	0.0	0.0	0.0	
	18:1;3/18:0;0	0.1	0.0	0.1	
	18:1;3/18:0;1	0.3	0.1	0.2	
	18:1;3/18:1;0		0.1	0.1	
	18:1;3/18:1;1		0.0	0.0	
*	18:1;3/20:0;0	0.2	0.0	0.1	
**	18:1;3/20:0;1	1.2	0.2	0.7	-2.9
	18:1;3/20:1;0	0.0	0.0	0.0	
	18:1;3/20:1;1	0.1	0.0	0.1	
	18:1;3/22:0;0	0.3	0.0	0.2	
**	18:1;3/22:0;1	13.6	1.5	7.5	-3.2
	18:1;3/22:1;0	0.1	0.0	0.0	
*	18:1;3/22:1;1	0.5	0.1	0.3	
*	18:1;3/24:0;0	0.5	0.1	0.3	
	18:1;3/24:0;1	3.4	4.6	4.0	0.4
	18:1;3/24:1;0	1.1	0.1	0.6	-3.0
**	18:1;3/24:1;1	48.6	9.8	29.2	-2.3
	18:1;3/26:0;0	0.1	0.1	0.1	
**	18:1;3/26:0;1	3.3	1.2	2.3	-1.5
	18:1;3/26:1;0	0.1	0.1	0.1	
**	18:1;3/26:1;1	3.2	0.9	2.1	-1.9
	18:1;3/28:0;0	0.1	0.0	0.1	
**	18:1;3/28:0;1	1.7	0.4	1.1	-2.2
	18:1;3/28:1;0	0.0	0.0	0.0	
*	18:1;3/28:1;1	0.3	0.1	0.2	

GIPC (Mol%)	L-mito
H-GIPC(18:0;3/24:1;1)	3.8
H-GIPC(18:1;3/16:0;1)	1.2
H-GIPC(18:1;3/22:0;1)	8.7
H-GIPC(18:1;3/24:0;1)	20.6
H-GIPC(18:1;3/24:1;1)	27.5
HN-GIPC(18:0;2/22:0;1)	8.3
HN-GIPC(18:0;2/22:0;1)	8.3
HN-GIPC(18:0;2/22:1;1)	3.8
HN-GIPC(18:0;3/22:0;0)	9.3
HN-GIPC(18:0;3/22:1;0)	5.2
HN-GIPC(18:1;3/22:0;0)	3.2

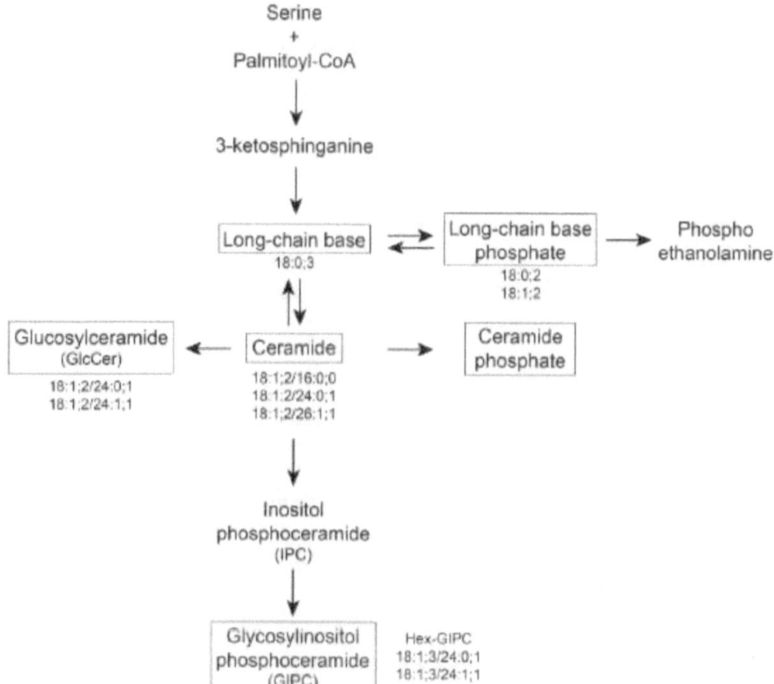

Figure 5. Profiles of molecular sphingolipid species, the according fold changes between L-mito and L-TE and general sphingolipid biosynthesis pathway in combination with mitochondrial enriched lipids.

Heat maps of (a) long chain bases, long chain base-phosphates and (b) complex sphingolipids illustrate the difference of species distribution based on LC-MS/MS analyses. Each block represents one lipid class wherein the detected species are presented and summarized to 100 %. Data represent mean values in mol % from three independent experiments. Binary logarithm was applied when the mean values are higher than 0.5 to inspect the fold changes (FC) between L-mito and L-TE. One and two asterisks (*, **) indicate p values < 0.05 and < 0.01, respectively, by Student's t-test. (c) General sphingolipid metabolism in *Arabidopsis*. Box frame: lipid classes analyzed by the LC-MS/MS approach with molecular lipid species enriched in L-mito in comparison to L-TE.

The common structure of sterols is a four-ring system, cyclopentanoperhydrophenanthrene, with possible conjugation of hydroxyl groups and acyl chains. In plants, a complex mixture of sterols can be found, including brassicasterol, campesterol, cholesterol, sitosterol and stigmasterol (Cacas *et al.* 2012). Campesterol was identified as major free sterol with 61.7 % in L-mito (Fig. 6). From the group of steryl glycosides cholesteryl and isofucosteryl glycoside accumulated preferentially in
L-

mito. Considering SE, 16:0 and 18:0 containing cholesteryl esters were enriched $2^{3.4}$ and $2^{5.3}$ folds, respectively, in L-mito. For ASG however the situation was blurred. DWF1 was the only protein related to sterol metabolism that we identified in plant mitochondria via proteomic analyses, but this protein was localized to ER membranes before that may be in close contact to mitochondria (Klahre *et al.* 1998). In summary, we observed an accumulation of GIPCs and free campesterol in leaf mitochondria.

Mitochondria harbor various additional metabolic pathways for the synthesis of lipophilic molecules. More than 20 fatty acid biosynthesis-related proteins were identified in our mitochondrial proteome (Fig. S5). They contribute to the synthesis of lipoic acid, ubiquinone and other terpenoid-quinones. Abundance, function and predicted localization of these proteins are specified in Tab. S1.

a

Free sterol (Mol%)	L-TE	L-mito	Mean	Log$_2$(FC)
** Campesterol		61,7	61,7	
Sitosterol	23,5	38,3	30,9	0,7
** Stigmasterol	76,5		76,5	

b

SG (Mol%)	L-TE	L-mito	Mean	Log$_2$(FC)
** Cholesteryl	0.2	2.2	1.2	3.2
** Sitosteryl	82.0	71.4	76.7	-0.2
Campesteryl	10.5	8.9	9.7	-0.2
Stigmasteryl	6.7	8.2	7.5	0.3
** Isofucosteryl		8.3	8.3	
Brassicasteryl	0.6	0.9	0.8	0.7

C

ASG (Mol%)

Sterol class	Sig		L-TE	L-mito	Mean	Log₂(FC)
Cholesteryl glycoside	*	16:0	0.1	1.3	0.7	3.9
		16:1	0.0	0.0	0.0	
		18:0	0.0	0.3	0.2	
		18:1	0.0	0.1	0.1	
		18:2	0.0	0.2	0.1	
		18:3	0.1	0.2	0.1	
	*	20:0	0.0	0.1	0.1	
	**	20:1	0.0	0.1	0.0	
	**	22:0		0.1	0.1	
	**	22:1	0.0	0.0	0.0	
Sitosteryl glycoside		16:0	36.3	40.2	38.2	0.1
		16:1	1.3	0.9	1.1	-0.5
		16:2	0.0	0.1	0.1	
	**	16:3	0.5	0.1	0.3	
	**	18:0	2.5	3.9	3.2	0.6
	**	18:1	6.9	4.1	5.5	-0.7
	**	18:2	22.9	13.0	18.0	-0.8
		18:3	8.8	10.6	9.7	0.3
		20:0	0.2	0.5	0.4	
		20:1	0.3	0.3	0.3	
	*	22:0	0.2	0.6	0.4	
		22:1	0.0	0.1	0.1	
	**	22:2	0.0	0.1	0.0	
Campesteryl glycoside		16:0	5.6	6.4	6.0	0.2
		16:1	0.1	0.1	0.1	
		18:0	0.4	0.5	0.5	0.4
		18:1	1.0	0.7	0.9	-0.5
		18:2	2.7	2.1	2.4	-0.4
		18:3	1.0	1.3	1.2	0.4
	*	20:0	0.0	0.1	0.1	
		20:1	0.1	0.0	0.1	
	*	22:0	0.0	0.1	0.1	
	**	22:1		0.0	0.0	
Stigmasteryl glycoside	**	16:0	3.0	6.8	4.9	1.2
		16:1	0.1	0.1	0.1	
	**	18:0	0.2	0.8	0.5	2.0
		18:1	0.7	0.4	0.6	-0.6
	*	18:2	0.8	1.1	0.9	0.5
	**	18:3	3.0	1.0	2.0	-1.7
	**	20:0	0.0	0.1	0.1	
		20:1	0.0	0.0	0.0	
		22:0	0.0	0.2	0.1	
		22:1	0.0	0.1	0.0	
Brassicasteryl glycoside		16:0	0.2	0.3	0.3	
		16:1	0.0	0.0	0.0	
	**	18:0	0.0	0.1	0.0	
		18:1	0.1	0.1	0.1	
		18:2	0.1	0.2	0.1	
	*	18:3	0.3	0.1	0.2	
		20:0	0.0	0.0	0.0	
		20:1		0.1	0.1	
		22:0	0.0	0.1	0.0	
		22:1	0.0	0.0	0.0	

SE (Mol%)

Sterol class	Sig		L-TE	L-mito	Mean	Log₂(FC)
Brassicasteryl ester		16:0	0.1	0.1	0.1	
		16:1	0.0	0.1	0.0	
		18:0	0.1	0.0	0.1	
	*	18:1	0.1	0.1	0.1	
		18:2	0.6	0.5	0.5	-0.2
		18:3	0.5	0.4	0.4	
	**	20:0	0.1		0.1	
		20:1	0.0		0.0	
	**	20:2	0.0		0.0	
	*	22:0	0.0		0.0	
		22:1	0.0		0.0	
		22:2	0.0		0.0	
Campesteryl ester		16:0	0.3	0.2	0.3	
		16:1	0.0	0.1	0.0	
		18:0	0.0	0.1	0.1	
	*	18:1	0.3	0.1	0.2	
		18:2	5.2	5.9	5.6	0.2
		18:3	4.4	5.2	4.8	0.2
		20:0	0.0		0.0	
	*	20:1	0.0		0.0	
	**	20:2	0.0		0.0	
	*	22:0	0.0		0.0	
		22:1	0.0		0.0	
Cholesteryl ester	**	16:0	0.2	1.9	1.0	3.4
	**	16:1	0.1	0.4	0.2	
	**	18:0	0.1	4.1	2.1	5.3
		18:1	0.3	0.6	0.4	
	*	18:2	0.8	2.6	1.7	1.8
		18:3	0.7	0.8	0.7	0.1
		20:0	0.0	0.1	0.1	
		20:1	0.0	0.0	0.0	
	*	20:2	0.0		0.0	
		22:0	0.1		0.1	
		22:1	0.1		0.1	
		22:2	0.0		0.0	
Sitosteryl ester	*	16:0	0.8	0.4	0.6	-1.1
		16:1	0.2	0.1	0.2	
		18:0	0.1	0.7	0.4	
	**	18:1	1.3	0.4	0.8	-1.7
	*	18:2	40.9	35.8	38.4	-0.2
		18:3	29.7	29.4	29.6	0.0
		20:0	0.1	0.0	0.1	
		20:1	0.2	0.0	0.1	
		20:2	0.0	0.1	0.0	
	**	22:0	0.5	0.1	0.3	
		22:1	0.0	0.0	0.0	
		22:2	0.0	0.0	0.0	
Stigmasteryl ester	*	16:0	0.3	0.2	0.2	
		16:1	0.0	0.1	0.1	
		18:0	0.1	0.1	0.1	
		18:1	0.3	0.2	0.2	
		18:2	6.1	4.8	5.4	-0.3
		18:3	4.7	4.5	4.6	0.0
		20:0	0.1		0.1	
		20:1	0.0		0.0	
	*	20:2	0.1		0.1	
	*	22:0	0.0		0.0	
	*	22:1	0.0		0.0	
		22:2	0.0		0.0	

Figure 6. Profiles of molecular sterol species, the according fold changes between L-mito and L-TE and general sterol biosynthesis pathway in combination of mitochondrial localized proteins and enriched lipids.

Heat maps of (a) free sterols (b) steryl glycosides (SG) and (c) acylated steryl glycosides (ASG) and steryl esters (SE) illustrate the difference of species distribution based on LC-MS/MS analyses. Each block represents one lipid class wherein the detected species are presented and summarized to 100 %. Data represent mean values in mol % from three independent experiments. Binary logarithm was applied when the mean values are higher than 0.5 to inspect the fold changes (FC) between L-mito and L-TE. One and two asterisks (*, **) indicate p values < 0.05 and < 0.01, respectively, by Student's t-test. (d) General sterol biosynthesis pathways in *Arabidopsis*. Heat maps visualize the protein abundance based on shotgun proteomic analyses of three independent experiments of purified mitochondria from leaves and cell culture. Boxed lipids: lipid classes analyzed by the LC-MS/MS approach with molecular lipid species enriched in L-mito in comparison to L-TE; green: proteins identified in the proteomic analysis of this study but predicted to localize in other organelles; magenta: proteins absent in the proteomic analysis of this study but predicted to localize in mitochondria. Full names and functions are itemized in Table S1. In the heat maps, proteins with high and low abundance are depicted in yellow and black, respectively. Predicted protein localization was based on The Arabidopsis Information Resource (TAIR; www.arabidopsis.org) and the Subcellular localization database for Arabidopsis proteins (SUBAcon; www.suba.live).

Lipid molecules can be imported by mitochondria

Transport of DGDG from plastids to mitochondria and reallocation of PE from IM to OM via the MTL complex have been demonstrated under phosphate-depleting conditions (Michaud *et al.* 2016). The MTL complex has more than 200 subunits and 11 have been verified to associate physically with the core subunits, Mic60 and Tom40, as previously shown by immunoblotting (Li *et al.* 2019, Michaud *et al.* 2016). In our mitochondrial proteome, 186 of the subunits (87 %) were identified including all the 11 verified ones (Fig. 7). The newly identified MTL subunit, DGS1, which links mitochondrial protein to lipid transport, is present in all mitochondrial extracts.

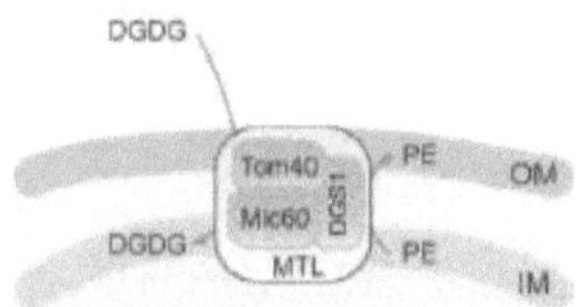

L-mito	C-mito	Locus	Name	Description
		AT1G06530	PMD2	Peroxisomal and mitochondrial division factor 2
		AT1G27390	TOM20-2	Translocase of outer membrane 20 kDa subunit 2
		AT2G37410	TIM17-2	Translocase inner membrane subunit 17-2
		AT2G42210	B14.7	Mitochondrial import inner membrane translocase subunit family protein
		AT3G20000	TOM40	Translocase of the outer mitochondrial membrane 40 kDa
		AT3G27080	TOM20-3	Translocase of outer membrane 20 kDa subunit 3
		AT4G39690	Mic60	Mitochondrial contact site and cristae organizing system subunit
		AT5G12290	DGS1	dgd1 suppressor 1
		AT5G13430	RISP	Ubiquinol-cytochrome c reductase iron-sulfur subunit
		AT5G40930	TOM20-4	Translocase of outer membrane 20 kDa subunit 4
		ATMG00160	COX2	Cytochrome oxidase 2

Figure 7. MTL complex-associated proteins in mitochondria.
Our proteomic analyses covered 186 of the 214 hypothetical subunits in the MTL complex; 11 have been investigated by immunoblotting approaches in previous studies. Heat maps visualize the protein abundance of three independent experiments of mitochondria purified from leaves and cell cultures. Proteins with high and low abundance are depicted in yellow and black, respectively.

In addition, at least 15 outer envelope (OE)-localized proteins from plastids were co-purified with mitochondria, with higher abundance in C-mito (Fig. 8). They account for 3.4 % (16/471) of the identified plastidial proteins, which are enriched by more than two folds in comparison to the proportion of OE proteins in plastids (46/3002, 1.5 %) (Inoue 2007, Inoue 2011, Kim *et al.* 2019, Simm *et al.* 2013). The enrichment of OE proteins in mitochondrial extracts suggests for a physical interaction between mitochondria and plastidial OE membrane. For instance, the membrane contact sites, which were not disrupted during the isolation procedure, and therefore are present in the mitochondrial samples as membrane patches.

L-mito	C-mito	Locus	Name	Function
		AT2G16640	TOC132	Translocon complex in the outer envelope membrane 132
		AT2G28900	OEP16-1	Outer plastid envelope protein 16-1
		AT2G43950	OEP37	Outer envelope protein 37
		AT3G06510	GGGT	Galactolipid:galactolipid galactosyltransferase
		AT3G17970	TOC64-III	Translocon at the outer membrane of chloroplasts 64-III
		AT3G46740	TOC75-III	Translocon at the outer envelope membrane of chloroplasts 75-III
		AT3G52420	OEP7	Outer envelope membrane protein 7
		AT4G02510	TOC159	Translocon at the outer envelope membrane of chloroplasts 159
		AT5G05000	TOC34	Translocon at the outer envelope membrane of chloroplasts 34
		AT5G19620	OEP80	Outer envelope protein of 80 kDa

Figure 8. Identified plastid outer envelope proteins in mitochondrial extracts.
Our proteomic analyses identified 12 outer envelope-localized proteins. Heat maps visualize the protein abundance of three independent experiments of mitochondria purified from leaves and cell cultures. Proteins with high and low abundance are depicted in yellow and black, respectively.

Discussion

In this study, we present an in-depth dataset of lipid molecular species from glycerolipids, sphingolipids and sterols of Arabidopsis leaf mitochondria. With the assistance of online resources, we assigned the lipid biosynthesis pathways within these mitochondria. Furthermore, we confirmed the existence of a protein complex for lipid trafficking between mitochondria and other organelles. In the past, the most abundant plant mitochondrial glycerolipid classes, such as PC, PE, PI, PA, CL, MGDG and DGDG, from *Arabidopsis* cell cultures and calli under phosphate-depleted conditions have been quantified with TLC-GC (Michaud *et al.* 2017). However, a TLC-GC-based approach lacks the molecular species information concerning acyl chain length and unsaturation degree of the lipids, which are critical parameters to determine the membranes physical properties. In addition, neither sphingolipids nor sterols of mitochondria were profiled to our very best knowledge formerly in *Arabidopsis*.

In our study, mitochondria were purified by differential and Percoll density gradient centrifugations. Much effort was made to document purity of the resulting organelle fraction: (i) visual inspection of the protein complex composition by 2D BN / SDS PAGE and (ii) label-free quantitative shot-gun proteins in combination with SUBAcon evaluation. The SUBAcon database integrates worldwide knowledge on subcellular localization information of Arabidopsis proteins based on *in vitro* or *in vivo* protein targeting experiments, mass spectrometry-based analyses of organellar fractions, protein-protein interaction data and bioinformatics tools for subcellular localization prediction. Based on the latter approach, purity of our mitochondrial fractions can be estimated to be in the range of 90% (87 to 94%, Fig. S2). Traces of chloroplasts were present in our fractions. However, even Rubisco, which is considered to be the most abundant chloroplast protein, was detectable only as a very faint spot on the 2D BN/SDS gel of the L-mito fraction (Fig. 1). Finally, quantitative lipidomics revealed the purity of our mitochondrial fraction. MGDG, DGDG, PG and SQDG are the main components of plastidial membranes (Hölzl and Dörmann 2019). In *Arabidopsis* chloroplasts, the ratio of MGDG : DGDG : PG : SQDG is about 1:0.5:0.1:0.3 (Awai *et al.* 2006). In contrast, a distinct ratio of 1:0.3:0.4:0 was detected in L-mito in this study (Fig. 2). Moreover, only a few specific MGDG and DGDG molecular species are enriched in L-mito compared to L-TE (Fig. 3B), suggesting that the cross-contamination from intact plastids or bulk plastidial membranes are insignificant. On the other hand, more than 85 % of the subunits of the MTL complex (Fig. 7) and many OE proteins (Fig. 8) were identified in the mitochondrial samples via our proteomics approach, strongly suggesting that a close contact between mitochondria and chloroplasts exists in our preparations. We conclude that the identified chloroplast lipids and proteins rather originate from small pieces of plastidial membranes and, importantly, the mitochondria-plastid contact sites, which are present in our mitochondrial fraction and contribute to the measured lipid composition. While the protein complexes connecting ER and mitochondria have been well described in yeast, their roles in mediating and/or facilitating lipid translocation are less defined in plants. In contrast to yeast, to our knowledge only one tethering protein has been described in plants up to now. The LEA-related LysM domain protein 1 (MELL1) is located at the junctions between mitochondria and ER in moss (Mueller and Reski 2015). The abundance of MELL1 correlated with the numbers of the mitochondria-ER contact sites, potentially involved in mitochondrial fusion and fission processes.

With the proteomic and lipidomic datasets as well as the online resources in hand, we assigned lipid biosynthesis pathways and ways of exchange of molecules between mitochondria, ER and plastids in plants in Fig. 9. Previous studies in yeast and mammalian cells have shown that mitochondria are capable of synthesizing PE, PA, PG and CL (Flis and Daum 2013, Horvath and Daum 2013, Tatsuta *et al.* 2014). Here, we expand the knowledge that glycerolipids including PE, PA, PG, and CL can be generated and/or modified in Arabidopsis leaf mitochondria. Considering PE biosynthesis, the rate-limiting enzymes in CDP-ethanolamine and PS decarboxylation pathways, PECT1 and PSD1, respectively, were identified in our mitochondrial samples with higher abundance in C-mito (Fig. 4a). This suggests that the generation of mitochondrial PE in the cell culture may be more active in comparison to the leaves. PE is one of the most abundant glycerophospholipids in biological membranes. Therefore, a high demand is expected to supply mitochondria for their biogenesis in actively dividing cell cultures. The other major component of biological membranes is PC. The last step of the PC biosynthesis pathway shares the same enzyme, AAPT, with PE synthesis. AAPT is an ER-localized enzyme and PC is thus considered to be synthesized in the ER and then transferred through MAM to mitochondria. The structural information of the enriched lipid species provides evidences for the biosynthesis in the ER as well. Beside those species carrying two C18 fatty acids on their glycerol backbone, many PE and PC species in mitochondrial samples have longer acyl moieties with 22 to 26 carbons, which can only be synthesized in the ER. Conventionally, the majority of PS biosynthesis was assumed to take place in the ER and before being transferred to mitochondria as well. PS and PE, although they are able to interconvert, have distinct lipid profiles from each other, suggesting that only selected lipid species are the substrates of PSD and PSS. A similar phenomenon applies to PI and phosphoinositides. That is, the most enriched PI species are not corresponding with the downstream PIP_2 species. Inositol polyphosphate phosphatase (SAL1), which generates inositol phosphates, was identified in our mitochondrial samples with higher abundance in C-mito comparing to L-mito. Inositol phosphates play crucial roles in many biological processes including gene expression and regulation of cell death through sphingolipids (Alcázar-Román and Wente 2008, Donahue *et al.* 2010). Removal of head groups from glycerophospholipids, mostly PC and PE, by phospholipase Dα1 (PLDα1) results in PA, serving as important precursor in other glycerophospholipid biosynthesis pathways. Comparing to L-TE, the most enriched PA species in L-mito carry 16:0 and 18:0 fatty acyl moieties. This corresponds to the enriched PI, PG and DAG species in L-mito, suggesting that PA may serve as an interconverting hinge between these glycerolipids in plant mitochondria.

Enzymes capable in synthesizing steps from PA to PG, including CDP-DAG synthases (CDS), PGP1, PGPP1, were found in our proteomic analysis and / or predicted to localize in mitochondria (Fig. 4a). Both PGP1 and PGPP1 have been identified in purified mitochondria and plastids, and we suggest that these two enzymes are present at the membrane junctions between these two organelles and are putative contact-site localized proteins. CL can be synthesized through condensation of PG and DAG by CLS, or addition of acyl chains to monolyso-CL by monolysocardiolipin acyltransferase (LCLAT). All identified enzymes involving in PG and CL biosynthesis, PGP1, CLS and LCLAT, are more abundant in C-mito compared to L-mito. This suggests an intensive interaction between mitochondria and plastids in cell cultures because (1) high amounts of proteins localized in contact sites imply a closer connection between the two organelles; (2) more CLS is required to convert the plastid-derived PG which is transferred to mitochondria presumably via membrane contact sites; (3) active energy generation is vital for actively dividing cells and thus a high amount of CL is required to assemble the respiratory chain complexes in mitochondria. Although PG is the precursor for CL synthesis, the abundance in L-mito is significantly less than in L-TE, probably because a substantial amount in L-TE originates from chloroplasts (Fig. 2).

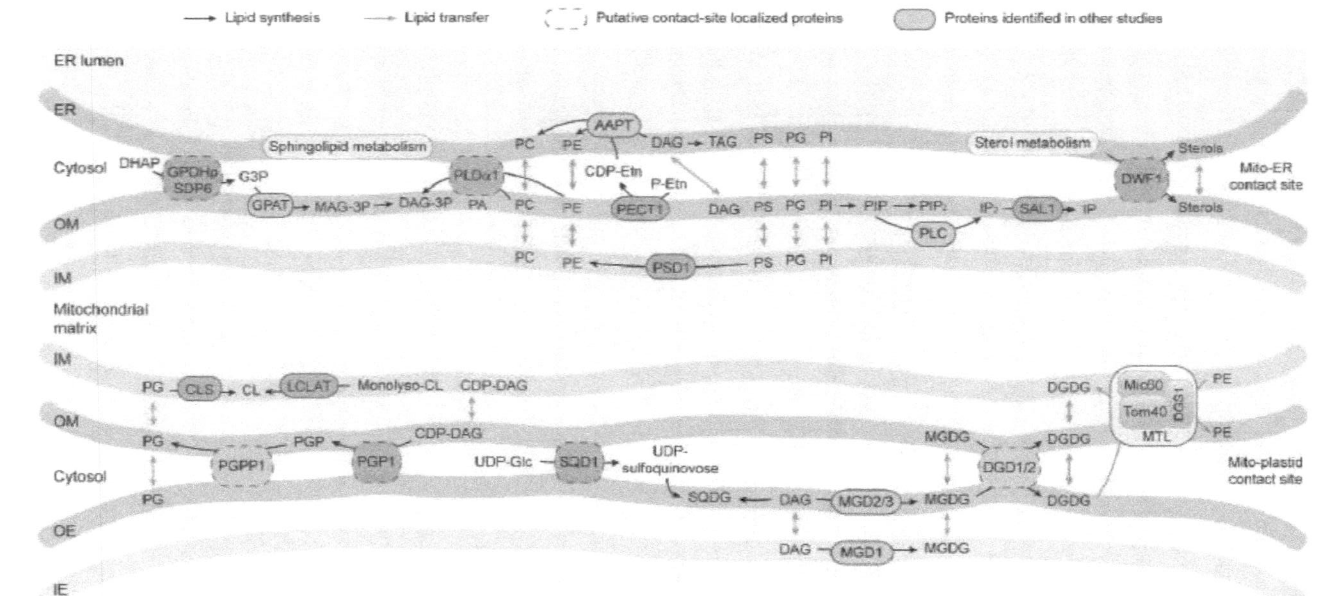

Figure 9. Model of lipid biosynthesis and trafficking within and between mitochondria, ER and plastids in *Arabidopsis*.

Lipid synthesis and transfer between membranes are indicated by black and blue arrows, respectively. Proteins identified in L-mito or C-mito with additional ER or plastidial localization based on The Arabidopsis Information Resource (TAIR; www.arabidopsis.org) and the Subcellular localization database for Arabidopsis proteins (SUBAcon; www.suba.live) are considered as putative contact-site localized proteins (dashed frame). Lipid biosynthesis-related proteins indicated in other studies are depicted in grey. Full names and functions of involved proteins are itemized in Table S1. OM, mitochondrial outer membrane; IM, mitochondrial inner membrane; OE, plastid outer envelope; IE, plastid inner envelope; MTL, mitochondrial transmembrane lipoprotein complex. DHAP, dihydroxyacetone phosphate; G3P, glyceraldehyde 3-phosphate; MAG, monoacylglycerol; DAG, diacylglycerol; Etn, ethanolamine; TAG, triacylglycerol; IP, inositol phosphate; PGP, phosphatidylglycerol phosphate; SQDG, sulfoquinovosyl diacylglycerol.

Typically, glyceroglycolipids such as MGDG, DGDG and SQDG are considered to be synthesized in plastids and are transferred to other organelles upon stress. For instance, DGDG is transferred from plastidial membranes to other compartments including mitochondrial membranes and the plasma membrane during phosphate starvation to compensate the loss of PC and PE (Jouhet *et al.* 2004). In addition to vesicular transportation and lipid trafficking via contact sites, emerging evidences support the hypothesis of lipid synthesis in *trans* in yeast and plants (Mehrshahi *et al.* 2013, Michaud *et al.* 2017, Tavassoli *et al.* 2013). That is, enzyme located at one membrane might be capable to catalyze the reaction on another membrane when they are in close apposition. In this way, neither tethering proteins nor massive lipid remolding under stress conditions is required, if the mitochondria can acquire lipids without setting up the junctions to other organelles. Therefore, we suspect that the DGDG biosynthesis enzymes, DGD1/2, catalyze the reactions in *trans* or at the mitochondria-plastid contact sites, providing DGDG molecules to compensate the loss of glycerophospholipids in real time (Fig. 4b). In addition, several subunits of the MTL complex were identified in our proteomic approach including Tom40, Mic60, DGS1 etc. (Fig. 7). It establishes a lipid trafficking system removing DGDG from plastids and allocating PE between the IM and OM. Rapid remodeling of the mitochondrial membranes is supported by in *trans* lipid biosynthesis and the MTL complex, likely in close cooperation with other complexes. However, the underlying mechanism of this lipid transportation machinery is still largely unknown. SQDG synthase (SQD1) was identified exclusively in C-mito, but no SQDG molecules were detected (Fig. 4b), suggesting that SQD1 is associated to the periphery of mitochondria as a result of the close apposition between mitochondria and chloroplasts in cell cultures.

Sphingolipid biosynthesis takes place at the ER membrane and subsequently in the Golgi apparatus for the addition of carbohydrate residues. Interestingly, in the last decade, several sphingolipid-metabolizing enzymes have been identified in mitochondria purified from yeast and mammalian systems, including Cer synthase and ceramidase (Bionda *et al.* 2004, Kitagaki *et al.* 2007, Novgorodov *et al.* 2011). In animal models, mitochondria-synthesized Cer plays a crucial role in cerebral ischemia-induced mitochondrial dysfunction. The connection between sphingolipids and mitochondria-promoted apoptosis has been proposed in plants as well. Recent studies in *Arabidopsis* have demonstrated that the ratio between LCB-P and Cer is involved in maintaining the balance of cell survival and apoptosis (Bi *et al.* 2014, Watanabe *et al.* 2013). However, no sphingolipid biosynthesis enzyme has been identified in plant mitochondria hereto (Fig. 5c). GIPCs are the most abundant sphingolipids in mitochondria, although still minor comparing to glycerophospholipids. It is known that PM-localized GIPCs are important in signal transduction and intercellular recognition (Ali *et al.* 2018, Lenarčič *et al.* 2017). GIPCs may as well establish the communication between mitochondria and other organelles, although further analysis is required to understand the functions of sphingolipids in plant mitochondria.

Sterols, although taking part in both biotic and abiotic stress responses, are at low abundance in mitochondrial samples. Sterol biosynthesis primarily takes place in the ER (Schaller 2003). Among them, sterol C-24 reductase (DWF1) was identified both in L-mito and C-mito, with higher abundance in C-mito. DWF1 has been proposed to mediate the biosynthesis of all phytosterols with higher specificity towards campesterol in Arabidopsis (Klahre *et al.* 1998, Youn *et al.* 2018). We suggest DWF1 as a putative contact-site localized protein according to the detailed species profile (Fig. 6). High amount of campesterol was detected in L-mito, suggesting the existence of an onsite biosynthesis and/or sterol transporter to facilitate the import of campesterol from the ER. However, unlike mammalian cells wherein cholesterol transport proteins such as steroidogenic acute regulatory protein, StAR (Clark *et al.* 1995), and MLN64 (Charman *et al.* 2010) have been identified, little is known about sterol transporters in plants.

In summary, we expanded the knowledge regarding lipid biosynthesis and modification in plant mitochondria by performing a global lipidome analysis of Arabidopsis leaf mitochondria to provide their in-depth lipid molecular species profile including glycerolipids, sphingolipids and sterols, suggesting that PE, PA, PS, PG, CL and free sterols can be synthesized in these organelles partially by the assistance of putative contact-site localized proteins and / or in *trans* lipid biosynthesis. Based on the proteomic results, we propose and confirm the existence of membrane contact site-localized proteins and their aspects in lipid biosynthesis pathways. This study serves as a foundation for additional researches in unveiling the functional roles of mitochondrial lipids and the mechanisms of mitochondria-dependent signaling pathways.

Materials and methods

Plant materials and growth conditions

Rosette leaves of wild-type *Arabidopsis thaliana* (L.) Heynh Columbia-0 were used for both extracting total lipid extract and purifying mitochondria. After sowing the seeds in pots with three-day cold stratification at 4 °C, seedlings were grown under 16 h-day length at 24 °C with 60 % relative humidity and 150 µmol photons m^{-2} sec^{-1} for one week. Young seedlings were transferred to large trays and grown with equal spacing for another three weeks before further experimental procedures.

Suspension culture of *Arabidopsis thaliana*

Suspension cultures were established starting from sterilized seeds of *Arabidopsis thaliana* wild-type Columbia-0, grown on MS medium plates containing 0.8 % agar. Plant pieces were transferred to B5 medium agar plates and cultivated for several weeks in the dark for callus induction. Callus was finally transferred into liquid B5 medium including 3 % (w/v) sucrose, 0.01 % (w/v) 2,4-dichlorophenoxyacetic acid and 0.001 % (w/v) kinetin. Cultures were incubated on a shaker at 24 °C in the dark. Callus was transferred weekly into new liquid medium (3 g / 100 ml).

Mitochondria isolation

From Arabidopsis rosettes

The mitochondria isolation procedure was described previously (Schikowsky *et al.* 2018). About 200 g of four-week-old *Arabidopsis* rosettes were collected and homogenized at 4 °C with a waring blender in 1 liter disruption buffer (0.3 M sucrose, 60 mM TES, 25 mM tetrasodium pyrophosphate, 10 mM potassium dihydrogen phosphate, 2 mM EDTA, 1 mM glycine, 1 % PVP40, 1 % BSA, 50 mM sodium ascorbate, 20 mM cysteine; pH 8.0) by three times for 10 sec with 30 sec intervals. The following procedures were performed on ice or at 4 °C. Two layers of miracloth with supporting gauze were used to filter the homogenate into a beaker. The remaining plant debris was first grinded with additional sea sand for 10 min by mortar and pestle, and then filtered again through miracloth. The filtrates were combined and centrifuged at 2,500 g for 5 min to eliminate the cell debris. Centrifugation with higher speed at 15,250 g for 15 min was applied on the supernatant to pellet mitochondria and other organelles. The resulting pellets were resuspended with a paintbrush in wash buffer (0.3 M sucrose, 10 mM TES, 10 mM potassium dihydrogen phosphate; pH 7.5). The samples were adjusted to the final volume of 12 ml with wash buffer and transferred to a Dounce homogenizer. Two strokes of pestle were performed to disrupt large organelles like chloroplasts. Aliquots of 1 ml samples were transferred carefully to Percoll gradients which had been established beforehand by 69,400 g centrifugation for 40 min in one-to-one ratio of Percoll and Percoll medium (0.6 M sucrose, 20 mM TES, 2 mM EDTA, 20 mM potassium dihydrogen phosphate, 2 mM glycine; pH 7.5). Mitochondria were separated from other components by centrifuging in the gradients at 17,400 g for 20 min. The resulting mitochondrial fractions formed white clouds at the bottom half of the gradients and were collected by Pasteur pipettes to clean centrifuge tubes. The clean-up procedures were performed three to five times by filling up wash buffer in the centrifuge tubes and pelleting the mitochondria with 17,200 g for 20 min, until the resulting pellet was firm. After each washing steps, two to three pellets were combined in one tube until all mitochondria from one biological replicate were pooled together. The mitochondrial pellets were weighed, resuspended with wash buffer and aliquoted at the concentration of 0.1 g/ml.

From cell cultures

Mitochondria isolation from *Arabidopsis thaliana* suspension cell culture was carried out as described before (Farhat *et al.* 2019). About 200 g fresh cells were harvested and homogenized using disruption buffer (450 mM sucrose, 15 mM MOPS, 1.5 mM EGTA, 0.6 % (w/v) PVP40, 2 % (w/v) BSA, 10 mM sodium ascorbate, 10 mM cysteine; pH 7.4) and a waring blender. During several washing steps, cell fragments were removed (centrifugation twice for 5 minutes at 2,700 g and once for 5 minutes at 8,300 g). Crude mitochondria were pelleted at 17,000 g for 10 minutes, resuspended in washing buffer (0.3 M sucrose, 10 mM MOPS, 1 mM EGTA; pH 7.2), homogenized using a Dounce homogenisator and loaded onto discontinuous Percoll gradients (phases of 18 %, 23 % and 40 % Percoll in gradient buffer (0.3 M sucrose, 10 mM MOPS, 0.2 mM EGTA; pH 7.2)). After ultracentrifugation (90 minutes, 70,000 g), purified mitochondria were collected from the 23 %-40 % interphase. For Percoll removal, several washing steps (10 minutes, 14,500 g) were performed using resuspension buffer (0.4 M mannitol, 1 mM EGTA, 10 mM tricine; pH 7.2) to gain a firm pellet of purified mitochondria.

Each of the three independently purified mitochondria populations from 4-week old rosette leaves (L-mito) and 200 g cell cultures (C-mito), respectively were used for all experiments.

BN/SDS-PAGE

Gel electrophoresis procedures (blue-native (BN) and SDS PAGE) were performed as described previously (Senkler *et al.* 2018), based on the published protocol given in (Wittig *et al.* 2006).

Label-free quantitative shotgun proteomics

Protein sample preparation

Sample preparation for shotgun proteome analysis was performed as described before (Thal *et al.* 2018). The protein content of the mitochondrial fractions was determined using a Bradford assay kit (Thermo Scientific, Rockford, USA). 50 µg protein of each sample were loaded onto a SDS gel for sample purification (Thal *et al.* 2018). Electrophoresis was stopped when the proteins reached the border between the stacking and the separating gel. Gels were subsequently incubated in fixation solution (15% (v.v) ethanol; 10% (v.v) acetic acid) for 30 min, stained for 1 h with Coomassie Brilliant Blue G250, and finally the protein band at the border of the two gel phases was cut out into cubes with edge lengths of approximately 1 mm. Trypsination of the proteins was carried out as described previously (Fromm *et al.* 2016).

Shotgun proteomic LC-MS analysis

Label-free quantitative mass spectrometric analyses of whole mitochondrial protein samples from cell culture and long day *Arabidopsis thaliana* leaves were performed as outlined before (Thal *et al.* 2018) using an Ultimate 3000 UPLC coupled to a Q Exactive Orbitrap mass spectrometer (Thermo Scientific, Dreieich, Germany).

Data processing

In a first step, the resulting MS data were processed using the Proteome Discoverer Software (Thermo Fisher Scientific, Dreieich, Germany) and searched with the Mascot search engine (www.matrixscience.com) against the tair10 protein database (downloaded from www.arabidopsis.org). For quantitative analyses, MS data were further processed as outlined before (Rugen *et al.* 2019) using the MaxQuant software package (version 1.6.4.0), the Andromeda search engine (Cox and Mann 2008) and the tair10 protein database. For determination of sample purity, peptide intensities were used, combined with the subcellular locations of assigned proteins as given by SUBAcon from the SUBA platform (www.suba.live) (Hooper *et al.* 2017). A proteomic heatmap was generated using the NOVA software (www.bioinformatik.uni-frankfurt.de/tools/nova/index.php)(Giese *et al.* 2015). Identified proteins of all six datasets (three biological replicates of leave total mitochondrial protein (L-mito 1,2,3) and cell culture total mitochondrial protein (C-mito 1,2,3)) were hierarchically clustered in a heatmap based on iBAQ (intensity based absolute quantification) values (for primary results see table S1).

Lipid extraction

Total lipid extract from leaf (L-TE) was obtained as described (Tarazona *et al.* 2015). Briefly, mixture of 150 mg frozen leaf powder or 100 mg mitochondria and 6 ml extraction buffer (propan-2-ol : hexane : water, 60:26:14 (v.v.v)) was incubated at 60 °C with shaking for 30 min. After centrifugation at 800 g for 20 min, the clear supernatant was transferred to clean tubes and evaporated under stream of nitrogen gas until dryness. Samples were reconstituted in 800 µl of TMW (tetrahydrofuran (THF) : methanol : water, 4:4:1 (v.v.v)). Procedure for extracting the mitochondrial lipids was adjusted by substituting the water fraction of the extraction buffer by equal volume of the mitochondrial aliquot. Further process was continued as described above.

TLC-GC quantification

TLC separation of lipid classes

Lipid extracts from 500 mg mitochondria and 50 mg leaves were spotted onto TLC 60 plates (20 × 20 cm², Merck KGaA, Darmstadt, Germany) in parallel with corresponding standards (Merck KGaA).

Extracts were developed by a solvent mixture of chloroform : methanol : acetic acid (65:25:8 (v.v.v)). After visualizing the lipid spots under 528 nm UV light, the bands were scrapped out and converted to fatty acid methyl esters (FAME) before GC analysis.

Acidic transesterification

Glycerolipids were transesterified by acidic methanolysis (Miquel and Browse 1992) and converted to fatty acid methyl esters (FAME). The lipid-bound silica powder from corresponding spots on TLC plate were scrapped out and added to 1 ml FAME solution (methanol : toluene : sulfuric acid : dimethoxypropane, 33:17:1.4:1 (v.v.v.v)) with 5 µg tripentadecanoin as an internal standard. After 1 h incubation at 80 °C, 1.5 ml saturated NaCl solution and 1.2 ml hexane were added subsequently. The resulting FAME was collected from the hexane phase, dried, and reconstituted in 10 µl acetonitrile.

GC/FID analysis

Lipid-bound fatty acids were analyzed after converting to FAMEs by a 6890N Network GC/FID System with a medium polar cyanopropyl DB-23 column (30 m × 250 µm × 25 nm; Agilent Technologies, Waldbronn, Germany) using helium as the carrier gas at 1 ml min^{-1}. Samples were injected at 220 °C with an Agilent 7683 Series injector in split mode. After 1 min at 150 °C, the oven temperature raised to 200 °C at the rate of 8 °C min^{-1}, increased to 250 °C in 2 min, and held at 250 °C for 6 min. Peak integration was performed using the GC ChemStation (Agilent Technologies).

Lipid derivatization

Phosphate-containing lipids - phosphatidic acids (PA), phosphoinositides (PIPs) and long-chain base phosphates (LCB-P) - were derivatized to enhance their chromatographic separation and mass spectrometric detection. Methylation procedure was applied on PA and PIPs as followed. Aliquots of 100 µl lipid extracts were first brought to dryness under stream of nitrogen gas and reconstituted with 200 µl methanol. Methylation reaction took place after the supply of 3.3 µl trimethylsilyldiazomethane. After 30 min incubation at room temperature, the reaction was terminated by neutralizing with 1 µl of 1.7 M acetic acid. Samples were dried under nitrogen gas and redissolved in 100 µl TMW. Acetylation procedure was applied on LCB-P. Aliquots of 100 µl lipid extracts were brought to dryness and reconstituted with 100 µl pyridine and 50 µl acetic anhydride. After 30 min incubation at 50 °C, samples were dried under stream of nitrogen gas with 50 °C water bath. To redissolve the samples, 100 µl TMW was used as the final solvent to proceed with lipid analysis.

Global lipidomic analysis with LC-MS

Analysis conditions and system setup were as described (Tarazona *et al.* 2015). Samples were separated by an ACQUITY UPLC system (Waters Crop., Milford, MA, USA) with a HSS T3 column (100 mm x 1 mm, 1.8 µl; Waters Crop.), ionized by a chip-based nanoelectrospray using TriVersa Nanomate (Advion BioScience, Ithaca, NY, USA) and analyzed by a 6500 QTRAP tandem mass spectrometer (AB Sciex, Framingham, MA, USA). Aliquots of 2 µl were injected and separated with a flow rate of 0.1 ml min^{-1}. The solvent system composed of methanol : 20 mM ammonium acetate (3:7 (v.v)) with 0.1 % (v.v) acetic acid (solvent A) and THF : methanol : 20 mM ammonium acetate (6:3:1 (v.v.v)) with 0.1 % (v.v) acetic acid (solvent B). According to the lipid classes, different linear gradients were applied: start from 40 %, 65 %, 80 % or 90 % B for 2 min; increase to 100 % B in 8 min; hold for 2 min and re-equilibrate to the initial conditions in 4 min. Starting condition of 40 % solvent B were utilized for long-chain bases (LCB) and phosphorylated long-chain bases (LCB-P); 80 % for diacylglycerol (DAG); 90 % for steryl esters (SE); 65 % for the remaining lipid classes. Retention time alignment and peak integration were performed with MultiQuant (AB Sciex). Quantitative results were calculated according to the amount of internal standards.

Biosynthesis pathways construction

The Arabidopsis Information Resource (TAIR; www.arabidopsis.org), the Subcellular localization database for Arabidopsis proteins (SUBAcon; www.suba.live), Kyoto Encyclopedia of Genes and Genomes (KEGG; www.genome.jp/kegg) and the shotgun proteomic analyses of L-mito and C-mito were combined to construct the lipid biosynthesis pathways in plant mitochondria. Additionally, lipidomics data are depicted in the pathways to illustrate the biosynthetic fluxes.

Acknowledgements

We are grateful for the technical assistance from Sabine Freitag. YL has been a doctoral student of the Ph.D. program "Molecular Biology" – International Max Planck Research School and the Göttingen Graduate School for Neurosciences, Biophysics, and Molecular Biosciences (GGNB) (DFG grant GSC 226) at the Georg August University Göttingen. IF and HPB acknowledge funding through the German Research Foundation (DFG: INST 186/822-1, INST 186/1167-1, INST 187/503-1 and ZUK 45/2010).

Supporting information
Table S1. Overall proteins identified in the proteomic approach.
Table S2. Overall lipids identified in the lipidomic approach.
Table S3. Lipid biosynthesis-related proteins identified in the proteomic approach.

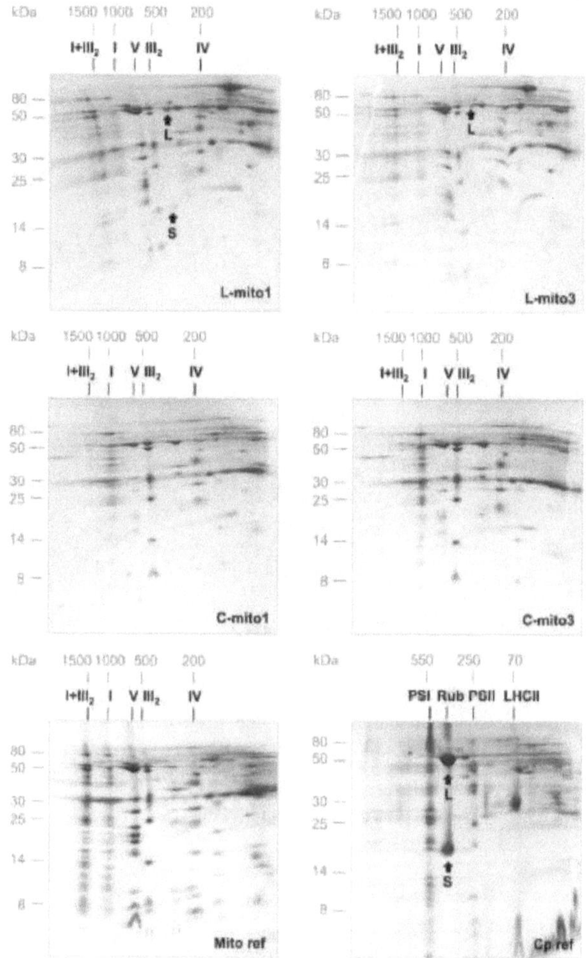

Figure S1. Purity inspection of mitochondrial protein complexes and supercomplexes by two-dimensional blue-native/SDS PAGE. Mitochondrial fractions isolated from Arabidopsis leaves (L-mito 1 and L-mito 3) and from Arabidopsis cell cultures (C-mito1 and C-mito 3) were separated by 2D PAGE and Coomassie-stained (corresponding gels of fractions L-mito 2 and C-mito 2 see Fig. 1). Numbers on top and to the left of the 2D gels refer to the masses of standard protein complexes / proteins (in kDa), the roman numbers above the gels to the identity of OXPHOS complexes (see Fig. 1 for detailed information). The arrows indicate the large (L; 53,5 kDa) and the small (S; 14.5 kDa) subunit of Rubisco. 2D blue-native/SDS reference gels for mitochondrial and chloroplast fractions (Mito-ref, Cp-ref) from *Arabidopsis thaliana* are given to the bottom of the figure (gels were taken from (Klodmann *et al.* 2011) and (Behrens *et al.* 2013)). Identity of the protein complexes visible on the chloroplast reference gel: PSI – photosystem I; PSII – photosystem II; Rub – Rubisco; LHCII – light harvesting complex II.

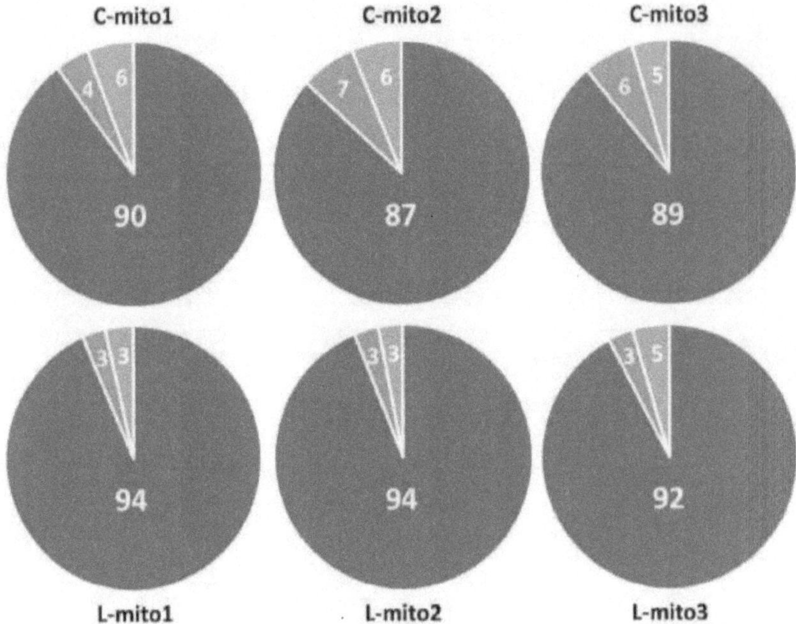

Figure S2. Purity of mitochondrial fractions as determined by label-free quantitative shot gun proteomics. Three mitochondrial fractions isolated from *Arabidopsis* leaves (L-mitos) and cell cultures (C-mitos) were analyzed. Summed-up peptide intensities were calculated for subcellular compartments based on protein assignments as given by the Subcellular localization database for Arabidopsis proteins (SUBAcon; www.suba.live). Blue: mitochondria; green: plastids; gray: others; numbers in %.

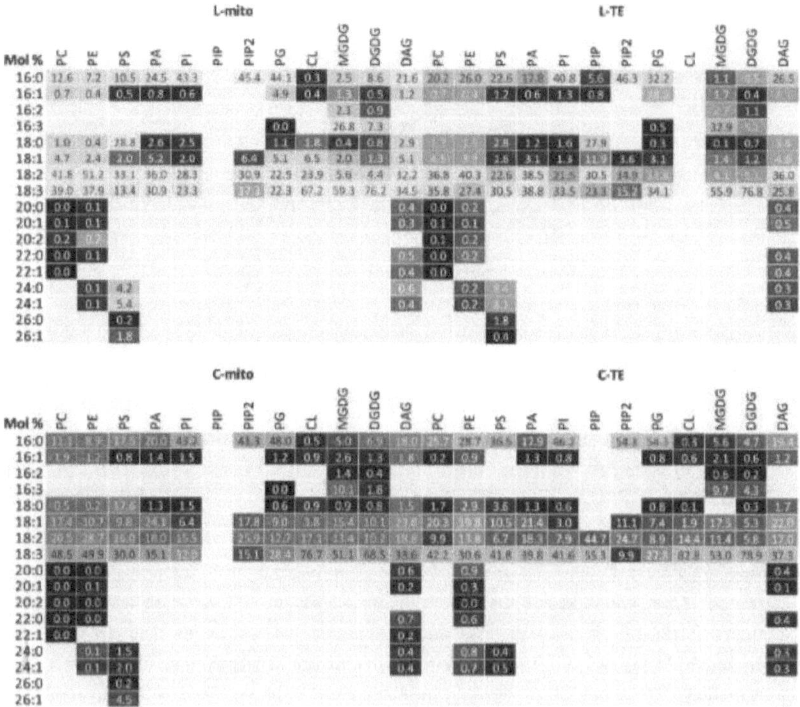

Figure S3. Fatty acid profiles of the glycerolipids from L-mito, L-TE, C-mito and C-TE. Heat maps illustrate the difference of the fatty acid distribution based on LC-MS/MS analyses. Each column represents one lipid class wherein the acyl moieties of all species are listed and summarized to 100 %. Data represent mean values in mol % from three independent experiments.

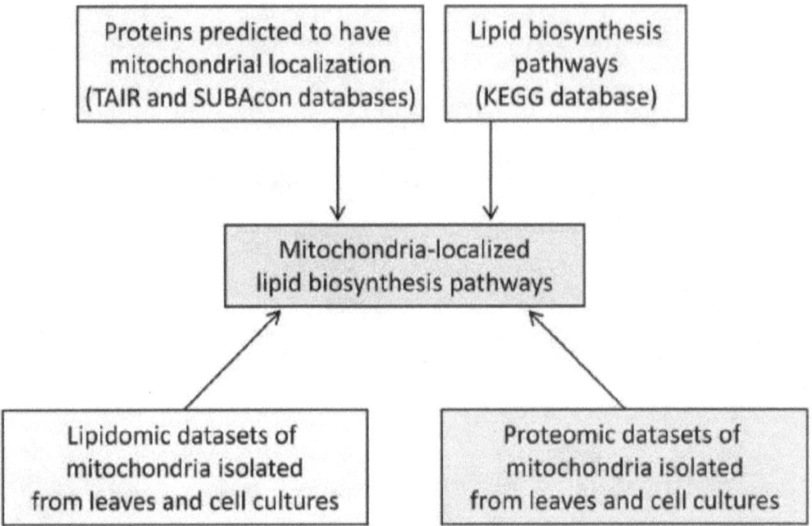

Figure S4. Workflow for the construction of the biosynthesis pathways. Multiple databases were combined to build the lipid biosynthesis pathways in mitochondria, The Arabidopsis Information Resource (TAIR; www.arabidopsis.org), the Subcellular localization database for Arabidopsis proteins (SUBAcon; www.suba.live), Kyoto Encyclopedia of Genes and Genomes (KEGG; www.genome.jp/kegg) and the proteomic datasets of isolated mitochondria in this study. Information of protein localizations and backbones of the biosynthesis pathways were obtained from TAIR, SUBAcon and KEGG, respectively.

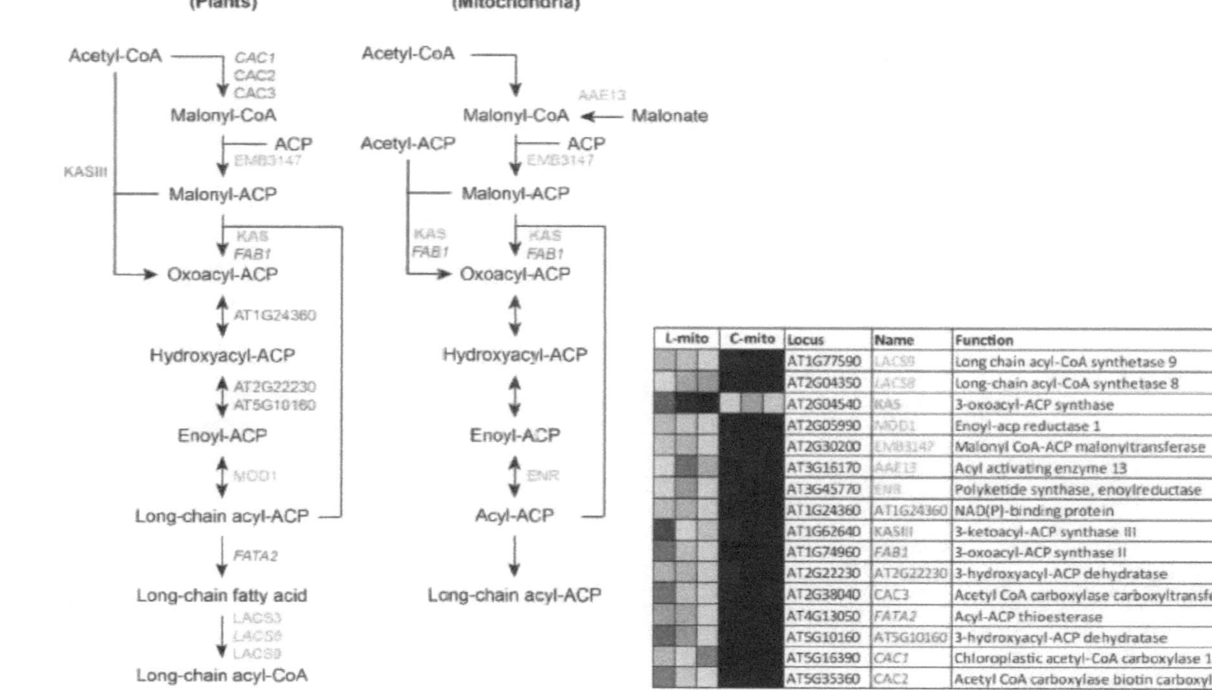

L-mito			C-mito			Locus	Name	Function
						AT1G77590	LACS9	Long chain acyl-CoA synthetase 9
						AT2G04350	LACS8	Long-chain acyl-CoA synthetase 8
						AT2G04540	KAS	3-oxoacyl-ACP synthase
						AT2G05990	MOD1	Enoyl-acp reductase 1
						AT2G30200	EMB3147	Malonyl CoA-ACP malonyltransferase
						AT3G16170	AAE13	Acyl activating enzyme 13
						AT3G45770	ENR	Polyketide synthase, enoylreductase
						AT1G24360	AT1G24360	NAD(P)-binding protein
						AT1G62640	KASIII	3-ketoacyl-ACP synthase III
						AT1G74960	FAB1	3-oxoacyl-ACP synthase II
						AT2G22230	AT2G22230	3-hydroxyacyl-ACP dehydratase
						AT2G38040	CAC3	Acetyl CoA carboxylase carboxyltransferase
						AT4G13050	FATA2	Acyl-ACP thioesterase
						AT5G10160	AT5G10160	3-hydroxyacyl-ACP dehydratase
						AT5G16390	CAC1	Chloroplastic acetyl-CoA carboxylase 1
						AT5G35360	CAC2	Acetyl CoA carboxylase biotin carboxylase

b Unsaturated fatty acid biosynthesis

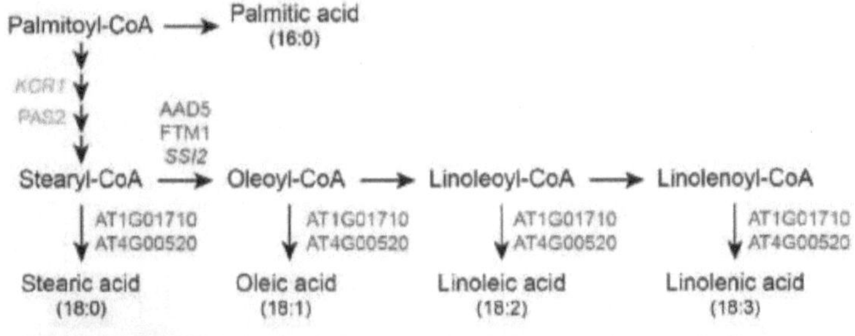

L-mito			C-mito	Locus	Name	Function
				AT1G67730	KCR1	Beta-ketoacyl reductase 1
				AT1G01710	AT1G01710	Acyl-CoA thioesterase II
				AT1G43800	FTM1	Stearoyl-acyl-carrier-protein desaturase
				AT2G43710	SSI2	Stearoyl-acyl-carrier-protein desaturase
				AT3G02630	AAD5	Stearoyl-acyl-carrier-protein desaturase
				AT4G00520	AT4G00520	Acyl-CoA thioesterase

C Lipoic acid metabolism

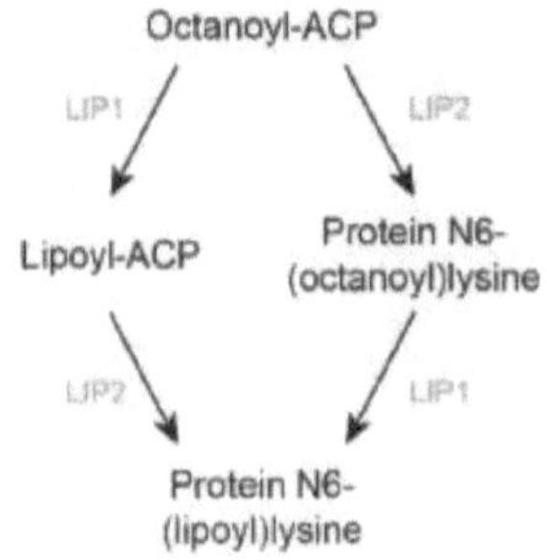

L-mito	C-mito	Locus	Name	Function
		AT1G04640	LIP2	Octanoyltransferase
		AT2G20860	LIP1	Lipoyl synthase

d Ubiquinone and other terpenoid-quinone biosynthesis

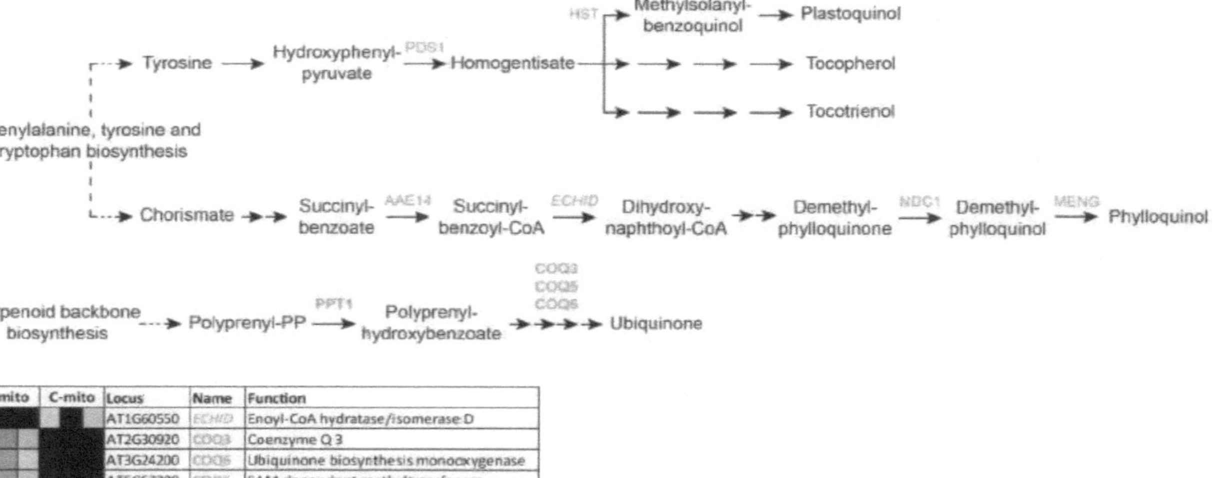

L-mito	C-mito	Locus	Name	Function
		AT1G60550	ECHID	Enoyl-CoA hydratase/isomerase D
		AT2G30920	COQ3	Coenzyme Q 3
		AT3G24200	COQ6	Ubiquinone biosynthesis monooxygenase
		AT5G57300	COQ5	SAM-dependent methyltransferase

Figure S5. Additional lipid biosynthesis pathways which do not contain analyzed lipid classes in this study. Pathways of (a) fatty acid biosynthesis, (b) unsaturated fatty acid biosynthesis, (c) lipoic acid metabolism and (d) ubiquinone and other terpenoid-quinone biosynthesis illustrate the general lipid biosynthesis in Arabidopsis. Heat maps visualize the protein abundance based on shotgun proteomic analyses of three independent experiments of purified mitochondria from leaves and cell culture. Blue: proteins identified in the proteomic analysis of this study and also predicted to localize in mitochondria; green: proteins identified in the proteomic analysis of this study but predicted to localize in other organelles; magenta: proteins absent in the proteomic analysis of this study but predicted to localize in mitochondria; bold font: exclusively localized in mitochondria; italic font: only identified in one of the mitochondrial populations. Further full names and functions are itemized in Table S1. In the heat maps, proteins with high and low abundance are depicted in yellow and black, respectively. Predicted protein localization was based on The Arabidopsis Information Resource (TAIR; www.arabidopsis.org) and the Subcellular localization database for Arabidopsis proteins (SUBAcon; www.suba.live).

Chapter 6. Discussions

The lipid composition of plant membranes is under tight control to ensure and maintain the cellular homeostasis and integrity, and it is challenged even further when plants are exposed to adverse growth conditions. Changes in temperature, nutrient availability or a disrupted lipid biosynthesis lead to highly specific variations of the lipid composition in the distinct membranes. The aim of this work was to gain conclusive insights into the special role of sphingolipids in membrane lipid remodeling at a subcellular level. First, the existing LC-MS-based lipidomics platform (Tarazona *et al.* 2015), which covers more than 300 molecular lipid species from over 20 different lipid classes, was enhanced further by expanding its coverage to include the minor but highly functional lipid classes such as complex glycosphingolipids, phosphorylated sphingolipids and phosphoinositides. In-depth analyses including lipidomics, proteomics and / or transversal lipid distribution studies were then applied on the *Arabidopsis* plasma membranes (PM) from non- and cold-acclimated wild type (WT) and sphingolipid mutants. In addition, a comprehensive study that combined lipidomics, proteomics and online databases mining of *Arabidopsis* mitochondria was performed. Altogether, this work contributed to reveal deeper insights into the functional roles of lipids in the organization and compositional dynamics of biological membranes.

6.1 Plant plasma membrane

6.1.1 The composition and organization of plant plasma membrane

The PM isolated from the leaves of *Arabidopsis* was profiled in detail with respect to the lipid species and lipid classes. The quantitative and qualitative analyses of glycerolipids present in PMs of WT demonstrated that lipid classes including PE, PS and PI are highly enriched in the PM in comparison to the TE. The contamination of plastidial lipids, SQDGs, are eliminated from the isolated PMs since they are only detectable in the TE, but not in the PM, via the highly sensitive LC-MS-based lipidomics platform (Fig. 2 and S1, Chapter 4). The molecular species profiles of all glycerolipid classes were revealed by the in-depth LC-MS lipidomics analysis. The most abundant species in PC, PE, PS, PI, phosphoinositides (PIP and PIP$_2$) and PA contain 16:0/18:2 and 16:0/18:3 moieties (Fig. 7, Chapter 4). Fatty acyl chains of these glycerophospholipids are conjugated to the *sn*-1 and *sn*-2 positions of the glycerol backbones by glycerol-3-phosphate acyltransferase (GPAT) and lysophosphatidic acid acyltransferase (LPAAT), respectively. Due to the distinct substrate specificities of the GPAT and LPAAT located in the plastids and the ER, which involved in the initial steps of prokaryotic and eukaryotic lipid biosynthesis pathways, respectively, characteristic signatures of the fatty acyl moieties can be identified in the resulting lipids. Namely, most of the lipids generated from the prokaryotic pathway (prokaryotic lipids) contain 18-carbon fatty acyl moieties and 16-carbon fatty acyl moieties at the *sn*-1 and *sn*-2 positions, respectively, whereas high amounts of lipids generated from the eukaryotic pathway (eukaryotic lipids) contain 16:0 at the *sn*-1 position and unsaturated 18-carbon fatty acyl moieties at the *sn*-2 position (Li-Beisson *et al.* 2013). Therefore, the predominant glycerophospholipids in the PM, which contain 16:0/18:2 and 16:0/18:3, are considered to be synthesized by the eukaryotic pathway. In addition, the eukaryotic lipids which contain very long-chain fatty acids (VLCFAs), especially PE and PS, are highly enriched in the PM. The biosynthesis of these eukaryotic lipids involves a series of reactions starting from the *de novo* biosynthesis of fatty acids in the plastids, followed by further desaturation, elongation and incorporation into lipids in the ER. Subsequently, these PM-enriched lipid species are specifically selected, however by an unknown mechanism, and sorted to the PM via the vesicular transport (Balla *et al.* 2020, Stefan *et al.* 2017). Besides the vesicular transport, extensive lipid trafficking between the PM and the plastids was identified by the detailed profiling of the typical plastidial lipid classes. The predominant MGDG and DGDG species in the PM contain 16:3/18:3 and 18:3/18:3 moieties, respectively, which corresponds to the observation in the TE (Fig. 8, Chapter 4). However, specific species such as MGDG (16:0/16:3) and DGDG (16:0/18:3) of eukaryotic origin were selectively enriched in the PM, indicating that the lipid trafficking between the PM and the plastids may be assisted by substrate-specific lipid transport proteins which can recognize the fatty acyl moieties. Noteworthy, it has been demonstrated that DGDG (16:0/18:3) accumulates under phosphate-limiting condition in the PMs (Andersson *et al.* 2003), which may result from the fact that the plants in this study were grown on soil that probably has lower phosphate availability comparing to the culture media. Only traces of the characteristic PG species of plastidial origin, PG (16:1/18:3), were detected in the PM whereas the eukaryotic PG (16:0/16:0) is highly enriched (Fig. 7, Chapter 4). This observation further emphasized that the lipid trafficking between the PM and the plastids is highly selective towards the transferred lipids, however via a sophisticated yet unknown mechanism.

The predominant sphingolipids in the PM of *Arabidopsis* leaves contain 18:1;3 as the LCB moieties and 24:0 or 24:1 as their fatty acyl chains (Fig. 3-5, Chapter 4). Noteworthy, comparing to ceramides Cers, which contain wide varieties of fatty acyl chains including 16:0, 22:0, 24:0, 24:1, 26:0 and 26:1, glucosylceramides GlcCers are conjugated preferentially with only 16:0, 22:0 and 24:1. This indicated that the GlcCer synthase in *Arabidopsis* may have higher substrate specificity towards these fatty acyl moieties. In addition, the molecular lipid species profiles of the GIPCs indicated that the glycosyl transferases, which produce GIPCs, possess distinct substrate affinities towards both the LCB backbones and the fatty acyl moieties. Namely, the hexosyl transferase, which generates hexosyl GIPCs (H-GIPCs), utilizes preferentially the substrates that contain an 18:1;3 LCB backbone with either a 24:0 or a 24:1 hydroxy fatty acyl moiety. In contrast, the *N*-acetylhexosaminyl transferase, which generates *N*-acetylhexosaminyl GIPCs (HN-GIPCs), utilizes preferentially substrates carrying an 18:0;3 LCB backbone and a 22:1 hydroxy fatty acyl moiety. However, the enzymes involved in the biosynthesis of GlcCers and GIPCs as well as their substrate specificities remain to be identified and experimentally characterized.

In *Arabidopsis* leaves, the sterols contain mostly campesterols or sitosterols as the core structures in both PM and TE (Fig. 6, Chapter 4). Campesterols and sitosterols both contain a hydroxyl group at the C3 position but differ in the side-chain modifications at the C24 position. Namely, campesterols contain a methyl residue and sitosterols contain an ethyl residue at their C24 position, so called 24-methylsterols and 24-ethylsterols, respectively. Noteworthy, it has been demonstrated that the ratio of 24-methyl- and 24-ethylsterols is specific for each plant species, and their balance plays significant roles in modulating the organization and permeability of biological membranes as well as the activity of the membrane protein including ion channels and signal transduction components (Clouse 2002, Valitova *et al.* 2016). In both PM and TE, sitosterol derivatives are the predominant ones in all sterol classes including sterol esters (SEs), steryl glycosides (SGs) and acylated steryl glycosides (ASGs), although campesterol derivatives consist of a substantial amount as well. This suggested that the ratio between 24-methyl- and 24-ethylsterols are maintained not only throughout the subcellular membranes but also in all the sterol classes of the *Arabidopsis* PM.

6.1.2 Membrane contact sites between plasma membrane and other organelles

Despite that the PM is involved extensively in the vesicular transport, by which it obtains the majority of its lipids, it communicates and exchanges lipids with other organelles also through the membrane contact sites. They are specialized microdomains built when two membranes come to close apposition. Several studies indicate that the PM can attach physically and establish membrane contact sites with the ER and the ER-PM contacts are involved in the endocytosis and autophagy pathways (Wang and Hussey 2019, Ye *et al.* 2020). It has been proposed that the formation and fusion of endosomes and autophagosomes occurs at the ER-PM contact sites before their vacuole internalization (Wang and Hussey 2019, Zhuang *et al.* 2016), however, the underlying mechanism is still elusive.

In *Arabidopsis*, a few ER-PM contact site localizing proteins have been described, including the synaptotagmins (SYTs) (Bayer *et al.* 2017, Pérez-Sancho *et al.* 2015), the vesicle-associated membrane protein–associated protein 27s (VAP27s), the VAP27-associated NETWORKED 3C (NET3C) (Rodriguez-Villalon *et al.* 2015, Wang *et al.* 2017) and the multiple C2 and transmembrane domain-containing protein (MCTP) (Grison *et al.* 2019). It has been demonstrated that these proteins and their mammalian orthologs not only tether the ER and the PM, but also interact directly with anionic lipids such as phosphoinositides and PS. The protein-lipid recognition and interaction may be essential in establishing and / or maintaining the structure of the ER-PM contact sites (Noack and Jaillais 2020). Noteworthy, in addition to the ER-PM interactions, the ER contacts diverse compartments in the cell including plastids, mitochondria, peroxisomes, lipid droplets and so on. Therefore, it has been hypothesized that the ER plays the central role in sorting the transferring lipids and coordinating the communications between all organelles (Andersson *et al.* 2007). Nevertheless, whether there are direct membrane contact sites between the PM and the other subcellular organelles without the involvement of the ER, as well as the modulation of the membrane composition by the ER require further investigations.

In this study, only specific lipids such as PG (16:0/16:0), MGDG (16:0/16:3) and DGDG (16:0/18:3) of plastid origin were selectively enriched in the PM, whereas other abundant plastidial lipids are depleted (Chapter 4). The results suggested that unknown substrate-specific lipid transport proteins may be involved in the lipid trafficking between PM and plastids. Alternatively, these lipids can be transferred first to the ER through the plastid-associated ER membranes (PLAMs) and be delivered subsequently to the PM through the ER-PM contact sites, by which an additional layer of modulation can be conducted. However, the current knowledge concerning the molecular mechanism of lipid trafficking and the lipid sorting signals (i.e. anionic lipids and / or lipid binding proteins) is still scarce.

6.1.3 The roles of sphingolipids in the organization of plant plasma membrane and their impacts on lipid modulation under cold stress

For plants, cold stress and pathogenic attack are two common environmental stresses that they encounter continually. To surpass the encountered stresses, it is critical for plants to adjust their PM lipid composition accordingly. For instance, increasing the unsaturation degrees of the abundant glycerophospholipids such as PC and PE are frequently used in many plant species to increase the membrane fluidity under cold stress (Tarazona *et al.* 2015, Uemura *et al.* 1995). In addition to glycerophospholipids, emerging evidences demonstrated that sphingolipids also play important roles in the adaptation of the PM lipid composition (Berkey *et al.* 2012, Huby *et al.* 2020, Michaelson *et al.* 2016). They are involved in numerous biological processes including pathogenic recognition and the establishment of the functional membrane microdomains, so called lipid rafts

(Laloi *et al.* 2007, Lenarčič *et al.* 2017). The roles of sphingolipids in the organization of the plant PM and their impacts on lipid modulation under cold stress are discussed here.

Under CA conditions, all complex sphingolipids including Cer, GlcCer, series A and B GIPCs are reduced in the WT PM (Fig. 11, Chapter 4). Diminished levels of glycosphingolipids (specifically GlcCer in most publications) have been commonly observed under CA conditions in many plant species, although the overall profile of their PM lipids vary greatly according to the cell type and developmental stages (Tab. 1). Decreased levels of glycosphingolipids were detected in leaves of oat, rye and potato when plants were grown under CA conditions for one to four weeks (Minami *et al.* 2008, Palta *et al.* 1993, Takahashi *et al.* 2016, Uemura *et al.* 1995, Uemura and Steponkus 1994). Noteworthy, it has been suggested that the cryotolerance of the PM can be enhanced by reducing the proportion of glycosphingolipids, which increases the hydration degree in the PM and therefore prevents the freeze-induced dehydration (Huby *et al.* 2020, Lynch and Dunn 2004, Webb *et al.* 1997).

Table 1. Levels of glycosphingolipids from plasma membranes isolated from leaves of NA and CA plants.

Plant species	NA (mol %)	CA (mol %)	Reference
A. thaliana			
A. thaliana Col-0[b]	2.9 ± 0.9	1.5 ± 0.6	This study[e, g]
A. thaliana Col-0[c]	7.3 ± 1.0	4.3 ± 1.4	Uemura *et al.* 1995[f, h]
A. thaliana Col-0[a]	4.4 ± 1.2	2.6 ± 1.1	Minami *et al.* 2008[f, h]
Oat			
A. sativa cv. N. almighty[d]	15.5 ± 0.7	12.6 ± 0.9	Takahashi *et al.* 2016[f, i]
A. sativa cv Ogle[d]	27.2 ± 1.0	24.2 ± 0.5	Uemura and Steponkus 1994[f, h]
A. sativa cv Kanota[d]	30.4	22.5	Uemura and Steponkus 1994[f, h]
Rye			
S. cereale cv. Maskateer[d]	13.4 ± 1.5	7.8 ± 1.8	Takahashi *et al.* 2016[f, i]
S. cereale cv Puma[d]	16.4 ± 1.0	10.5 ± 0.5	Uemura and Steponkus 1994[f, h]
Potato			
S. commersonii[b]	6.1 ± 0.5	4.9 ± 0.8	Palta *et al.* 1993[f, i]
S. tuberosum[b]	6.5 ± 1.0	5.0 ± 0.6	Palta *et al.* 1993[f, i]

Plants were acclimated in cold for [a]7 days, [b]10 days, [c]2 weeks or [d]4 weeks.
The data represent the levels of [e]all glycosphingolipids or [f]GlcCer specifically.
Lipid quantification was performed via [g]LC-MS, [h]TLC-GC or [i]spectrophotometric analyses.

The in-depth lipidomics analysis of sphingolipids revealed that hydroxylated Cer, GlcCer and H-GIPC (hCer, hGlcCer and hH-GIPC, respectively), which contain 18:1;3 LCB backbone with 24:1 fatty acyl moiety, increase significantly under CA conditions in *Arabidopsis* leaves (Fig. 3-4, Chapter 4). This observation correlates well with previous studies, which demonstrated that higher levels of GlcCer with 18:1;3 LCB backbone (Imai *et al.* 1997, Imai *et al.* 2000, Kawaguchi *et al.* 2000) and monounsaturated hydroxylated fatty acyl chains (predominantly 24:1) (Imai *et al.* 1995) are present in chilling-resistant plants. Interestingly, analysis of the *Arabidopsis* sphingolipid Δ^8 LCB desaturase mutant (*sld1sld2*), which is impaired in introducing double bonds at the Δ^8 position of the LCBs, displayed a reduced level of LCB (18:1;3) and a higher sensitivity towards cold stress in comparison to the WT (Chen *et al.* 2012, Zhou *et al.* 2016b). Therefore, it has been suggested that it is the Δ^8 unsaturated LCBs (i.e. LCB (18:1;3)) rather than the Δ^8 unsaturated LCB-conjugated GlcCer that increases the cryotolerance (Chen *et al.* 2012). Nevertheless, the alterations in both the LCB backbone and the fatty acyl moiety of the sphingolipids with respect to hydroxylation and desaturation have been demonstrated to be involved in the lateral lipid – lipid interactions in the PM, which influence the hydration degree and modulate the membrane cryobehaviors (Steponkus *et al.* 1990, Webb *et al.* 1997).

In an attempt to address the functions of specifically the hydroxylated sphingolipids and the effect of their absence on the modulation of the PM lipid composition upon cold stress, the *Arabidopsis* sphingolipid biosynthesis mutant with impaired sphingolipid fatty acid α-hydroxylases (*fah1 fah2*) was subjected to CA condition in this work. Reduced levels of α-hydroxylated sphingolipids, which is characteristic in *fah1 fah2* plants can be observed in the lipid species profiles of both TE and PM (Fig. 4, Chapter 4). This phenomenon is especially prominent in the GlcCer species, which results in the reduced level of total GlcCer in the *fah1 fah2* plants (Fig. 10, Chapter 4). Similar to GlcCer, strong reductions of series A GIPCs (H-GIPC and HN-GIPC) were observed in both TE and PM of *fah1 fah2*, which was even more prominent in the PM. Therefore, it is proposed that the biosynthesis of both GlcCer and series A GIPCs as well as their transport in *fah1 fah2* plants are impaired due to the loss of the α-hydroxyl groups on the complex sphingolipids. Interestingly, although the overall Cer levels are reduced proportionally in the *fah1 fah2* PM as well, they are increased in the *fah1 fah2* TE in comparison to the WT TE. This indicates that the increased proportion of Cer in *fah1 fah2* plants resides primarily at the intracellular membranes, presumably at the ER, rather than being transported to the PM. They may structurally compensate the loss of other hydroxylated sphingolipids in the intracellular membranes, or the hydroxylation process may serve as the checkpoint for the transport of the Cers to the PM.

The lipid profile of the *fah1 fah2* PM isolated from plants grown under NA condition was highly reminiscent to that of the WT PM from plants grown under CA condition (Fig. 10-11, Chapter 4).

The significant reductions of complex glycosphingolipids including GlcCer and series A GIPCs in the WT PM under CA condition, which have been described to increase the hydration and cryostability of the PM (Huby *et al.* 2020, Lynch and Dunn 2004, Webb *et al.* 1997), were also detected in the lipid profile of the *fah1 fah2* PM under NA condition. This suggests that the loss of α-hydroxylated sphingolipids and cold stress may trigger similar responses in remodeling the lipid composition of the *Arabidopsis* PM. Moreover, not only the reduction of complex sphingolipids, but also the variations in selected glycerolipids and sterols display similar patterns. Namely, lower levels of PC, PS, DAG, SG and ASG, higher levels of PA and SE. One of the few exceptions concerns the relative amount of LCB-P. Previous studies have demonstrated that cold stress triggers the accumulation of LCB-Ps and stimulates the expression of cold responsive genes in *Arabidopsis* cell cultures and seedlings (Ali *et al.* 2018, Cantrel *et al.* 2011, Dutilleul *et al.* 2012, Guillas *et al.* 2013). However, a reduced level of LCB-Ps was observed in the WT PM under CA condition (Fig. 11, Chapter 4). This suggests that the accumulation of LCB-Ps stimulated by cold stress may be restricted to intracellular membranes, presumably the ER and the Golgi apparatus where the respective LCB kinase is located (Crowther and Lynch 1997, Zäuner *et al.* 2010). On the other hand, higher LCB-P levels were observed in the lipid profile of the *fah1 fah2* PM under NA condition (Fig. 10, Chapter 4). Especially the increase of LCB-P (18:1;3) may structurally compensate the loss of α-hydroxylated and glycosylated complex sphingolipids by their trihydroxylated and phosphorylated residues, respectively. Thereby, the elevated level of LCB-P (18:1;3) may contribute to maintain the physical property and the membrane stability of the *fah1 fah2* PM. Moreover, the loss of complex glycosphingolipids including GlcCer and series A GIPCs in the *fah1 fah2* PM under NA condition may be rescued by the elevated levels of series B GIPC as well. The glycan residues of series B GIPCs may interact with other multi-glycosylated molecules such as glycoproteins and surface glycan to form glycoconjugates and thereby increase the membrane stability of the *fah1 fah2* PM (Bucior and Burger 2004, Handa and Hakomori 2017, Popescu *et al.* 2003).

Another explanation that the *fah1 fah2* PM under NA condition resembles the WT PM under CA condition can be accessed from the formation and the structural components of the lipid rafts. Complex sphingolipids and sterols have been identified to be critical structural components of plant lipid rafts (Borner *et al.* 2005, Cacas *et al.* 2016, Laloi *et al.* 2007, Mongrand *et al.* 2004, Simons and Ikonen 1997, Takahashi *et al.* 2016). Reduced levels of the lipid raft-forming lipids including Cer, GlcCer, H-GIPC and HN-GIPC as well as SG and ASG were observed from both lipid profiles of the *fah1 fah2* PM under NA condition and the WT PM under CA condition (Fig. 10-11, Chapter 4). The lack of the structural components of lipid rafts suggests a limited capacity in lipid raft formation within the *fah1 fah2* PM. Noteworthy, it has been demonstrated that the PMs of *fah1 fah2* protoplasts are less orderly packed in comparison to the WT (Lenarčič *et al.* 2017). In addition, the ordered and lipid raft-like microdomains on the PM, the DRMs, are involved extensively in the cold-induced responses. A few specific proteins in the DRMs such as P-type H$^+$-ATPases, synaptotagmin homolog SYT1 and endocytosis-related proteins have been identified to increase significantly under CA conditions, (Minami *et al.* 2009, Takahashi *et al.* 2013, Yamazaki *et al.* 2008). They play important regulatory roles in different biological processes under CA condition. Namely, the PM H$^+$-ATPase, (Ishikawa and Yoshida 1985, Martz *et al.* 2006), has been demonstrated to be involved in sphingolipid desaturation (Borner *et al.* 2005, Chen *et al.* 2012). SYT1 is related to calcium-dependent PM resealing of the freezing tolerance mechanism (Takahashi *et al.* 2013, Yamazaki *et al.* 2008). Finally, the inhibition of endocytosis as well as the disassembly of microtubules have been associated with the CA process in plants (Abdrakhamanova *et al.* 2003, Bolte *et al.* 2004). Collectively, the characterization of DRM-localizing proteins demonstrates that the structure and the functions of DRMs, and probably also of lipid rafts, are involved in the cold-induced responses in plants. This further supports the observation that the loss of lipid raft-forming lipids triggers similar mechanism as cold acclimation in modulating the organization of the *Arabidopsis* PM.

The interrelation of sphingolipids and cold stress may be linked further to other biological processes such as the phytohormone alterations and pathogenic responses (Huby *et al.* 2020). It has been demonstrated that the AtFAH (especially ALFAH1)-interacting protein, the ER-localized cell death suppressor Bax inhibitor-1 (AtBI-1), is involved in the plants' reactions to various biotic and abiotic stresses, including pathogenic responses, SA-triggered cell death, ER stress and oxidative stress (Ishikawa *et al.* 2009, Kawai-Yamada *et al.* 2009, Kawai-Yamada *et al.* 2004, Watanabe and Lam 2008). In addition, AtBI-1 regulates sphingolipid biosynthesis by interacting with other sphingolipid modification enzymes including sphingolipid base hydroxylase 2 (AtSBH2), acyl lipid desaturase 2 (AtADS2) and sphingolipid Δ^8 LCB desaturase 1 (AtSLD1) via cytochrome b_5 (Nagano *et al.* 2014, Nagano *et al.* 2012). In rice (*Oryza sativa*), overexpression of BI-1 leads to an enrichment of α-hydroxylated GlcCer in isolated DRMs as well as to the loss of DRM-localized proteins that are involved in the SA- and oxidative stress-triggered cell death such as Flotillin Homolog (FLOT) and Hypersensitive-Induced Reaction Protein 3 (HIR3) (Ishikawa *et al.* 2015, Nagano *et al.* 2016). Noteworthy, constitutively elevate levels of SA have been detected in the *fah1 fah2 Arabidopsis* plants with increased resistance against the obligate biotrophic fungus, *Golovinomyces cichoracearum* (König *et al.* 2012); however, corresponding *fah1 fah2* plants from rice contain similar SA level as WT plants and are more susceptible to the hemibiotrophic fungus, *Magnaporthe oryzae* (Nagano *et al.* 2016). Nevertheless, these studies indicate that sphingolipid biosynthesis proteins regulate the lipid and protein composition of lipid rafts, which harbor proteins involved in

SA signaling pathways. In addition, SA-triggered responses have been demonstrated to associate with cold stress as well, since enhanced levels of SA, particularity of glycosylated derivatives, have been detected in several plant species under CA conditions (Kosová *et al.* 2012, Scott *et al.* 2004, Wan *et al.* 2009). Therefore, SA signaling may indeed associate with the modulation of lipid composition and thus contribute to the similar lipid profiles as observed in the *fah1 fah2* PM at NA condition and the WT PM at CA condition. The relative profiles of lipid class and lipid species generated by the LC-MS-based lipidomics workflow in this work provide invaluable information concerning the modulation of membrane organization in the plant PM under cold stress. Building on this work, investigation via absolute quantitative approaches may allow further insights into the underlying mechanisms of membrane remodeling and the regulatory functions of specific lipid species.

6.1.4 Asymmetric distribution of plasma membrane lipids

The components of the PM are distributed asymmetrically across the membrane bilayer, as well as within each monolayer. As introduced in Chapter 1, our knowledge concerning the transversal distribution of the PM lipids arises from the studies of mammalian erythrocytes that most of the glycosphingolipids and PC are localized in the outer leaflet while PE, PS, PI, PIP and PIP_2 are localized in the inner leaflet (Bretscher 1972, Devaux 1991, Devaux and Morris 2004, Di Paolo and De Camilli 2006, Harayama and Riezman 2018, Lorent *et al.* 2020, Verkleij *et al.* 1973). Compared to the animal system, only very few studies have focused on the transversal distribution of the PM lipids in plants. This animal membrane-based prototype has been proposed to globally represent the PM lipid distribution including the plant PM (Gronnier *et al.* 2018, Jaillais and Ott 2020); however, only to certain extent.

In this work, highly selective lipid analyses were combined with the in-depth lipidomics approach to elucidate the transversal distribution of the most abundant glycolipids and glycerophospholipids within the PM isolated from WT *Arabidopsis* leaves. The major GlcCer species are distributed in a species-depend manner to the different monolayers of the PM (Fig. 12a, Chapter4). About 60 % of the most abundant GlcCer molecular species, hGlcCer (18:1;3/24:1), were detected in the apoplastic leaflet. The preferential distribution towards the apoplastic leaflet can be detected for other GlcCer species as well. We have used periodate to experimentally address the distribution of GIPCs by the same approach as utilized for GlcCer. That is, the exterior leaflets of each of the two PM vesicle populations, bearing either the apoplastic-side-out or the cytoplasmic-side-out orientations, were treated chemically by periodate to derivatize the sugar headgroups of the glycolipids. However, the recovery rate of periodate-treated GIPCs was too low and no convincing signal was detected in the LC-MS-based analysis. To optimize the recovery rate, the parameters of the centrifuging steps including the composition of the buffer system and the centrifugation speed need to be fine-tuned further. It should be noted that the feasibility of enzymatically digesting the GIPCs by sphingolipid ceramide *N*-deacylase (SCDase) has been evaluated as well. However, the GIPC analogue, ganglioside GM1 that serves as the standard for GIPC in this work, cannot be hydrolyzed in the tested conditions. A negative result from a similar enzymatic experiment with SCDase was recorded in a previous publication as well (Cacas *et al.* 2016). Nevertheless, endogenous GIPC is presumed to be distributed within the apoplastic leaflet of the PM according to (1) its biosynthesis which generates lipid molecules in the inner leaflet of Golgi apparatus, followed by subsequent transport via Golgi-derived vesicles and exposure to the cell surface, (2) its bulk hydrophilic headgroups that presumably hinders a transmembrane movement and (3) its biological function that is associated with pathogenic recognition on the cell surface (Cacas *et al.* 2016, Lenarčič *et al.* 2017).

Plants contain a wide variety of sterols and sterol derivatives, of which the transversal distribution cannot be assessed in the animal system. In-depth lipid profiling revealed that the distribution of the most abundant SG and ASG species differs significantly (Fig. 12b, Chapter 4). For instance, sitosteryl glycoside is found exclusively within the cytoplasmic leaflet, whereas the transversal distribution of the acylated sitosteryl glycosides varies greatly among individual molecular species. As SG and ASG are structural components of lipid rafts together with the GlcCer and GIPC (Ferrer *et al.* 2017, Grosjean *et al.* 2015, Roche *et al.* 2008), specific sterol species (such as 16:0-sitosteryl glycosides) may have higher affinity towards these complex sphingolipids and are thus allocated preferentially to the apoplastic leaflet. On the other hand, the ASG species with unsaturated fatty acyl moieties (18:2 and 18:3) reside predominantly in the cytoplasmic leaflet, which may contribute to a previous finding that the cytoplasmic leaflet of the eukaryotic PM displays a lower viscosity, due to the high unsaturation degrees of the localized glycerophospholipids (Lorent *et al.* 2020). Interestingly, it has been observed in oat that under phosphate starvation, ASG and DGDG are used to replace phospholipids (especially PC) at the apoplastic and the cytoplasmic side of the PM, respectively, to provide phosphate for other necessary biological processes (Tjellström *et al.* 2010). This suggests that the distribution of ASGs across the PM and the apoplastic leaflet may be part of the responses triggered by certain abiotic stresses.

In comparison to sphingolipids and sterols, the distribution of glycerophospholipids has been investigated more thoroughly in the literature. Here, the transversal distribution of PE, PC, PS and

PI was addressed at the level of molecular lipid species (Fig. 13, Chapter 4). PC, PE and PI species are distributed nearly symmetrically across the membrane bilayer of the *Arabidopsis* PM with minor variations. This corresponds to a previous observation that glycerophospholipids are distributed equivalently on both leaflets of the PM isolated from hypocotyl cells of mung beans (Takeda and Kasamo 2001). Noteworthy, we observed that the majority of PS is found in the cytoplasmic leaflet. Together with other anionic lipids such as PA and phosphoinositides, it has been proposed to establish the electrostatic signature of the PM to recruit proteins targeted to lipid rafts (Colin and Jaillais 2020, Furlan *et al.* 2020, Jaillais and Ott 2020, Noack and Jaillais 2017, Platre *et al.* 2018, Simon *et al.* 2016). For instance, studies have indicated that the GTPase Rho of Plants 6 (ROP6) is stabilized by PS in the lipid rafts and can modulate auxin signaling locally during root development (Kay and Fairn 2019, Platre *et al.* 2019). Albeit PS has been suggested to be actively kept in the cytoplasmic leaflet by flippases (Cacas *et al.* 2016, Devaux and Morris 2004), we also detected some in the apoplastic leaflet as reported in a previous publication (O'Brien *et al.* 1997). These results correlate well to the former knowledge at the level of lipid class, with additional insights into the distribution of lipid molecular species. Clearly, further investigations are required to understand the function of individual lipid species on each monolayer of the plant PM and the modulation of the lipid distribution under different environmental stresses.

The approach to address the membrane asymmetry in this study was constructed based on the lipid class-specific treatments on the selected leaflet and the subsequent lipid analysis. Namely, the apoplastic- and cytoplasmic-side-out PM vesicles were treated with either periodate or phospholipase A_2, which specifically derivatizes the glycolipids or digests the glycerophospholipids, respectively, that are present on the outer leaflet. Consequently, the lipidome of each PM monolayer was calculated based on the reduced signals in the treated outer leaflet of the two PM vesicle populations with opposite membrane orientation. Both reagents were selected based on their specificities towards a wide range of substrates as well as their commercial availability. Although a few studies have performed a similar workflow, however, no molecular information of plant lipids was provided due to their subsequent lipid class or lipid category-dedicated analyses (Lynch and Phinney 1995, Takeda and Kasamo 2001, Tjellström *et al.* 2010). Alternative approaches utilizing lipid-specific protein probes have demonstrated the distribution of minor anionic lipid classes such as PS and PIPs via flow cytometry or fluorescence microscopy (O'Brien *et al.* 1997, Platre *et al.* 2018). Moreover, sterols and sterol derivatives can also be sensed by protein probes such as Filipin III, which binds to the 3-OH residue on the core structure of the sterols and can be analyzed via immunoblotting (Tjellström *et al.* 2010). Although the distribution of lipid class or lipid categories can be visualized by the photometric-based techniques or immunoassays, it is challenging to obtain quantitative results as well as the information of the lipid molecular species. On the other hand, LC-MS analysis enables the investigation of the molecular species information of not only the abundant lipid species but also the minor and functional ones. However, further development in the LC-MS analysis is required to achieve absolute quantification of individual lipid species present on the PM.

6.2 Mitochondria

6.2.1 Mitochondrial contact site-localizing proteins in plants and their roles in lipid metabolism

Mitochondria isolated from leaves and cell cultures of *Arabidopsis* were comprehensively investigated in Chapter 5, in an attempt to determine their capacity to synthesize lipids and to reconstruct the lipid metabolism pathways in plant mitochondria. The results led to the conclusion that many phospholipid classes including PE, PA, PS, PG and cardiolipins (CL) as well as FS can be synthesized or at least be further modified in mitochondria, partially by the assistance of lipid biosynthesis proteins localized at the membrane contact sites between the mitochondria and other intracellular organelles as has been described recently (Michaud *et al.* 2017). It has been demonstrated that the mitochondrion obtains lipids through the non-vesicular transportation pathway. That is, unlike the PM, which obtains its lipids predominantly via vesicular transport, the mitochondrion obtains a substantial amount of its lipids through membrane contact sites (Galmes *et al.* 2016, Michaud *et al.* 2016). Increasing evidences have revealed that membrane contact sites play important roles in not only lipid transport, but also carry several other biological functions including coordinating metabolic activities, facilitating signaling pathways, mediating organelles fission and organizing compartments within multimembrane organelles like mitochondria and plastids (Friedman *et al.* 2011, Hoyer *et al.* 2018, Prinz *et al.* 2020, Wong *et al.* 2019).

Several membrane contact site-localizing proteins were proposed based on the proteomics and lipidomics analyses presented in Chapter 5. Namely, the phosphatidylglycerophosphate synthase 1 (PGP1), the phosphatidylglycerophosphate phosphatase 1 (PGPP1), the DGDG synthases (DGD1) and the sterol C-24 reductase (DWF1). PGP and PGPP are essential enzymes in the biosynthesis of PG, which in turn is the precursor of mitochondria-specific lipids, CL (Fig. 4a, Chapter 5). In addition, both enzymes have been identified to localize in the inner mitochondrial membrane (IM) of yeast and mammalian systems (Džugasová *et al.* 1998, Horvath and Daum 2013, Kawasaki *et al.* 2001, Osman *et al.* 2010, Vance 2015, Zhang *et al.* 2011), in plastids (Hölzl and Dörmann 2019) and in mitochondria of *Arabidopsis* (Tanoue *et al.* 2014, Wada and Murata 2009). PG is present in high amounts in both plastids and mitochondria in *Arabidopsis* and its synthesizing enzyme PGP1 has been identified to carry the targeting presequence of both organelles as well (Babiychuk *et al.* 2003). The plant PGP and PGPP were therefore proposed to localize in or closely to mitochondria, providing PG molecules for CL biosynthesis in proximity. Nevertheless, mitochondria may obtain PG also from the ER, probably through the mitochondria-associated ER membranes (MAMs). It should be noted that CLs can also be generated by lysocardiolipin acyltransferase (LCLAT or Tafazzin), which is involved extensively in the remodeling of CLs under oxidative stress in mammalian systems (Cao *et al.* 2004). However, further characterization of LCLAT is required to understand its function in planta and its contribution in the remodeling of plant mitochondrial membranes. Noteworthy, it has been suggested that lipid molecules can be involved in the tethering between intracellular membrane systems as well. In yeast, a tethering protein between mitochondria and the PM, Num1, has been characterized to interact directly with phosphoinositides on the PM by its pleckstrin homology (PH) domain and with CL on mitochondria by its N-terminal coiled-coil domain (Ping *et al.* 2016). Nevertheless, whether it is conserved in plants that lipid molecules are involved in initiating, establishing and stabilizing the membrane contacts site requires further investigations.

The DGDG biosynthesis enzyme, DGD1, has been identified both in mitochondrial and plastidial membranes, although DGDG presents predominantly in plastids under optimal growth conditions (Fig. 4b, Chapter 5). Nevertheless, upon phosphate starvation, a high amount of DGDG is transferred rapidly from the plastids to the mitochondria (Michaud *et al.* 2016), suggesting that DGD1/2 may closely associate to or physically interact with the mitochondrial transmembrane lipoprotein (MTL) complex, which imports DGDG into the mitochondria from the plastids.

The sterol biosynthesis enzyme, DWF1, is proposed to localize at the membrane contact sites between the mitochondria and the ER. Although AtDWF1 has been visualized to locate at the ER (Klahre *et al.* 1998), it is identified in the mitochondrial proteome based on LC-MS analysis. In addition, significantly higher amounts of its products, campesterols, were detected in the mitochondria compared to total lipid extract via lipidomics analysis (Fig. 6, Chapter 5), suggesting an onsite biosynthesis at the plant mitochondria and/or the presence of campesterol-specific transporters to facilitate the import of campesterols from the ER. However, unlike the mammalian systems wherein sterol transporters like the steroidogenic acute regulatory protein, StAR (Clark *et al.* 1995), and MLN64 (Charman *et al.* 2010) have already been characterized, the sterol transporters in plants remain to be identified.

In plants, several evidences have indicated that mitochondria can establish membrane contact sites with the ER, plastids, peroxisomes, vacuoles and the PM. However, only two protein systems localized at the membrane contact sites between mitochondria – plastids and mitochondria – ER, respectively, have been described and characterized hereto. In *Arabidopsis*, a large lipid-enriched complex, the MTL complex, has been identified to have the capability of exporting PE from mitochondria and importing DGDG from plastids to mitochondria (Michaud *et al.* 2016) under phosphate starvation (Härtel *et al.* 2000, Jouhet *et al.* 2004, Moellering and Benning 2011). In addition, characterization of MTL has revealed that it contains over 200 subunits, and many of them are components of the translocase of the outer membrane (TOM) such as Tom40, and the mitochondrial contact site and cristae organizing system (MICOS) such as Mic60 (Li *et al.* 2019, Michaud *et al.* 2016). Therefore, it has been postulated that Mic60, the central component of MTL complex, can regulate the tethering between the IM and outer mitochondrial membrane (OM) and initiate the lipid transport (Michaud *et al.* 2016, Michaud and Jouhet 2019). Another characterized membrane contact site-localizing protein is the mitochondria-ER-localized LEA-related LysM domain protein 1 (MELL1) in moss (*Physcomitrella patens*). It has been identified to tether mitochondria and the ER and to be involved in several processes including mitochondrial constriction and fission (Mueller and Reski 2015). However, whether MELL1 is conserved in higher plants and involved in the lipid transfer between the ER and mitochondria are still elusive.

Plant mitochondria can establish membrane contact sites with multiorganelles simultaneously as well. For instance, three-way junctions between mitochondria, peroxisomes and chloroplasts have been identified under radical-inducing conditions such as high light intensity and photorespiration (Hanson and Hines 2018, Jaipargas *et al.* 2016, Mathur *et al.* 2012, Oikawa *et al.* 2015, Pérez-Sancho *et al.* 2016, Shai *et al.* 2016, Sinclair *et al.* 2009). Similar three-way junction established by mitochondria-early endosome-ER has been observed as well (Hsu *et al.* 2018). However, the physiological functions of these junctions between multiorganelles are still largely unknown. Although several researches based on yeast and mammalian cells have suggested that membrane contact sites are involved not only in lipid transport, but also in regulating the lipid-based signaling pathways, organizing the membrane structure and membrane genesis (Liu and Li 2019, Prinz *et al.* 2020), the operating mechanisms and the biological functions of membrane contact sites in plants remain to be investigated.

Noteworthy, emerging evidences contribute to the novel hypothesis that lipids can be synthesized in *trans* in yeast and plants (Mehrshahi *et al.* 2013, Michaud *et al.* 2017, Tavassoli *et al.* 2013). That is, the enzymes localized in one membrane can catalyze the biosynthesis reaction on another membrane when two membranes are in proximity. By means of in *trans* lipid biosynthesis, neither tethering proteins nor massive lipid remolding under stress conditions, such as the rapid influx of DGDG into mitochondrial membranes under phosphate starvation condition, is required. Nevertheless, cautious inspections should be conducted to elucidate its underlying mechanism and the coordination of in *trans* lipid biosynthesis with other lipid transport pathways.

In this study, two omics analyses, namely lipidomics and proteomics, have been integrated and cross-compared intensively in order to extract conclusive information, combined with the online databases and used to construct the lipid biosynthesis pathways in Arabidopsis leaf mitochondria. The obtaining results expanded the previous knowledge which was proposed based on the mitochondrial lipid compositions of the Arabidopsis pale tissues (i.e. cell cultures and calli) and the protein subcellular localization from published resources (Michaud et al. 2017). Although some steps in the glycerolipid biosynthesis pathway in plant mitochondria have been denoted accordingly, a few additional involving proteins, especially the ones in the metabolism of phosphoinositides, were identified via the conducted combinatorial approach (Chapter 5). Unlike the widespread strategies that combine omics datasets from genomics, transcriptomics, proteomics and / or metabolomics, which have several established online tools and commercially available software (Eicher *et al.* 2020), the integration of lipidomics into a multi-omics approach is much less developed. Therefore, large numbers of comparisons and statistical analyses were required to integrate the lipidomics and proteomics analyses in this study. However, requirement for routine implementation of lipidomics with other omics data will likely occur and accelerate the development of an efficient platform in the near future. Here, several putative membrane contact site-localizing proteins between mitochondria and other organelles were proposed as well through the multi-omics analysis, although further validations by biochemical characterization are required, the flow and combination of information between omics studies of lipids and proteins has been demonstrated.

6.2.2 The functional role of sphingolipids in mitochondria

None of the sphingolipid biosynthesis enzymes have been identified to localize in mitochondria in plants, and sphingolipids are presumably transferred through membrane contact sites to mitochondria after being synthesized in the ER and the Golgi apparatus (Fugio *et al.* 2020). Sphingolipids are considered to be important membrane components of mitochondrial membranes as well and can be identified in the lipid profiles of the mitochondria isolated from *Arabidopsis* leaves (Chapter 5). When compared to the lipid profile from the overall leaf extract of *Arabidopsis*, mitochondria contain significant lower proportions of sphingolipids with 18:1;3 LCB backbone (Fig. 5, Chapter 5), which is the predominant sphingolipid species in the TE. This suggests that a separate pool of sphingolipids is present in the mitochondria and may exert specific functions locally as demonstrated in animal cells (Birbes *et al.* 2001, Colombini 2010).

One of the essential biological processes governed by mitochondria is PCD). It is associated with the cell development, defense reactions against pathogens and responses towards environmental stresses (Hirsch *et al.* 1998, Jacoby *et al.* 2012). It has been demonstrated that mitochondria initiate PCD by producing and releasing bioactive molecules as cell-death signals. For instance, cytochrome *c*, which locates in the intermembrane space between the IM and OM of mitochondrial membranes, is released to the cytoplasm from mitochondria as a PCD signal (Basova *et al.* 2007, Rodrigues *et al.* 2007). In addition, reactive oxygen species (ROS), which are produced by the mitochondria, can alter the permeability of mitochondrial membranes, and excess amount can initiate the PCD (Hirsch *et al.* 1998). Noteworthy, sphingolipids have been indicated to regulate PCD as well in yeast, animal cells and plants (Berkey *et al.* 2012). Several studies have indicated the functions of sphingolipids on the mitochondria-induced PCD in mammalian cells; however, whether sphingolipids coordinate with mitochondria in plants and the underlying mechanism that regulates the PCD require further investigations.

Among sphingolipid classes, Cer is the most investigated sphingolipid class in connection with mitochondrial functions in both plants and mammalian systems. It has been indicated that Cer accumulates in the *Arabidopsis orm* mutant, which has an impaired function of orosomucoid protein resulting in enhanced activity of the serine palmitoyltransferase (SPT) and stimulated *de novo* sphingolipid biosynthesis (Li *et al.* 2016). In addition, the mutant plants display an early-senescence phenotype with ROS production in mitochondria and at the cell wall. A similar result was obtained from the *Arabidopsis acd5* mutant, which lacks the ceramide kinase (CERK) activity. The *acd5* mutant contains elevated levels of Cer and accumulates ROS in mitochondria, and shows spontaneous PCD and activated autophagy (Bi *et al.* 2014). Moreover, the ACD5 protein has been identified in several subcellular organelles, including the ER, the PM, the Golgi apparatus and the mitochondria. As a result, it has been proposed that CERK localizing at different compartments can exert different functions; that is, CERK at the PM may be involved in the defense responses at the cell surface while CERK in mitochondria may suppress the mitochondria-mediated cell death signaling. Since the function of CERK is to generate phosphorylated Cer (Cer-P), detailed analysis of Cer-P in biological extracts can contribute to the knowledge regarding the functional roles of Cer-P at different subcellular localizations and in the regulation of mitochondria-mediated PCD. The endogenous amount of Cer-Ps in mitochondria was not sufficient to be analyzed via the lipidomics workflow alone. Nevertheless, in combination with the approach presented in Chapter 3, which incorporates the profiling of endogenous Cer-P in the LC-MS-based lipidomics workflow after TLC purification, creates a great potential in obtaining further insights. Interestingly, treatment with exogenous Cer can also stimulate PCD in mitochondria. For instance, it has been demonstrated that direct treatment of *Arabidopsis* protoplasts with Cer carrying 2- or 6-carbon fatty acyl moieties (C2- and C6-Cer), which are permeable through the cellular membrane, can induce the release of cytochrome *c* and initiate PCD (Yao *et al.* 2004). This strongly corroborates the correlation of sphingolipids with the mitochondria-mediated stress responses. However, a recent study has revealed that the release of cytochrome *c* does not occur in rice (*Oryza sativa*) protoplasts after exogenous Cer treatment (Zhang *et al.* 2020), indicating that sphingolipids may exert specific regulatory functions in different plant species.

In animal systems, the mechanisms of Cer-mediated mitochondria dysfunction have been investigated in more detail (Fugio *et al.* 2020). Interesting results have been obtained by investigating the functions of Cers containing 16- and 18-carbon fatty acyl moieties (C16-Cer and C18-Cer, respectively) in connection to cell death responses. For instance, C16-Cer has been described to inhibit the activity of mitochondrial complex IV and stimulate the production of ROS, thereby provoking oxidative stress (Zigdon *et al.* 2013). On the other hand, instead of increasing ROS production, C18-Cer has been demonstrated to mediate the autophagic cell death and mitophagy by direct interaction with mitochondria-targeting proteins (Oleinik *et al.* 2019, Sentelle *et al.* 2012). This suggests that the acyl chain length of the Cer molecules is critical in determining its functions in mitochondria such as the initiation of mitophagy. Furthermore, it has been demonstrated that only Cer molecules generated by mitochondria-localized sphingomyelinase, which degrade sphingomyelin to generate Cer and phosphocholine, are involved in the activation of apoptosis (Birbes *et al.* 2001). In addition, Cer molecules that accumulate at the mitochondria, due to the mutation in a Cer transport protein (CERT) (Rao *et al.* 2014), create channel-like structures at the OM of mitochondria and thus lead to mitochondria dysfunction (Chang *et al.* 2015, Nielson and Rutter 2018, Ueda 2015). Altogether, these results indicate that, in addition to the acyl chain length of the Cer molecular species, the site of biosynthesis and the subcellular localization of Cer can regulate the functions of mitochondria as well. Nevertheless, whether these phenomena are conserved in plants and whether they are involved in the stress responses towards biotic and / or abiotic stresses require further investigations.

6.3 Membrane fractionation is required to broaden the detection range and achieve global lipidomics analysis

As presented in Chapter 4 and 5, the enhanced lipidomics platform provided comprehensive insights into the lipid composition of the PM and the mitochondria. This enhanced lipidomics platform includes the minor lipid classes such as complex glycosphingolipids, phosphorylated sphingolipids and phosphoinositides, which are critical in numerous biological processes. The methodological development targeting these minor lipid classes was building on the establishment of subcellular membrane fractionation, including the PM, the mitochondria and the MC which are the crude endomembrane fractions. Since lipids are fundamental components of biological membranes and different subcellular membranes are composed of distinct lipid compositions, the isolation of specific subcellular membranes can be considered as a purification procedure for the minor lipid classes. In this work, the purification of the PM, the mitochondria and the MC has contributed to optimize the LC-MS parameters for the pre-existing detection methods of SQDG and H-GIPC; and to establish the methods and identify PIP, PIP_2, CL, LCB-P, Cer-P, HN-GIPC and dihexosyl GIPC (H-H-GIPC). Chemical derivatization has also enhanced the mass spectrometric detection of several lipid classes, and was incorporated as part of the standard workflow; namely, methylation was applied on PIP and PIP_2 and acetylation on LCB-P and Cer-P before the LC-MS analyses.

Additional lipid classes such as complex glycolipids with tri- or tetragalactosyl headgroups, which are widely distributed in algae and lower plants, are not yet included in this enhanced plant lipids-orientated lipidomics platform. Nevertheless, a wide coverage of the lipid classes, also the minor ones, belonging to glycerolipids, sphingolipids and sterols were successfully integrated into the enhanced lipidomics platform. Compared to the conventional approaches for lipid analysis, which are mostly lipid class-specific, the information of lipid molecular species can be addressed by the presented lipidomics platform. This enables future detailed characterization of plant membrane lipids, including the minor lipid classes, and contributes to the understanding of the functional role exerted by individual lipid species.

6.4 Concluding remarks

This work has greatly expanded the knowledge about the lipid landscape of two distinct biological membranes in plant cells; namely, the plasma membrane (Chapter 4) and the mitochondria (Chapter 5). With the asset of the enhanced wide-ranging LC-MS-based lipidomics workflow (Chapter 2) and further advancement focusing on the bioactive minor lipids, Cer-P (Chapter 3), the membrane organization and the functions of specific lipids have been addressed.

The compositional remodeling of the PM lipid molecular species under CA condition and the impacts of the loss of the α-hydroxylated sphingolipids in *Arabidopsis* were addressed in detail (Chapter 4). The result presented in this chapter led to the conclusion that the loss of sphingolipid α-hydroxylases triggers similar responses as CA in the PM, with respect to the modulation of the PM composition at the level of lipid class, which affects the membrane organization and the lipid-mediated signaling pathways. In addition, the loss of sphingolipid α-hydroxylases does not diminish the fact that lipid molecular species with higher desaturation degrees accumulate under cold acclimation, indicating that the fine-tuning of the membrane property occurs at the level of lipid species. Furthermore, species-specific transversal distribution of the most abundant phospholipids and glycolipids reveals an additional sophistication in modulating the membrane organization. The profiling of lipid species further assists the frontier research on distinct functions that individual lipid species may exert.

The combinational approach with lipidomics, proteomics and online databases mining further defines the lipid biosynthesis pathways and possible lipid biosynthesis capacity in plant mitochondria (Chapter 5). Lipid classes including PE, PA, PS, PG, CL and FS can be synthesized or modified in plant mitochondria with partial assistance of putative contact site-localized proteins and / or *trans* lipid biosynthesis. The presented comprehensive dataset facilitates further investigations on identifying and characterizing the membrane contact site-localizing proteins between mitochondria and other organelles. Moreover, the enhanced capacity of the lipidomics workflow (Chapter 2 and 3) creates a great potential in obtaining further insights into the functions of specific lipid class and lipid species in biological membranes, such as sphingolipid-mediated cell death in mitochondria.